T0135813

Multiphysical Modelling of Regular and Irregular Combustion in Spark Ignition Engines using an Integrated / Interactive Flamelet Approach

Von der Fakultät
für Umweltwissenschaften und Verfahrenstechnik
der Brandenburgischen Technischen Universität Cottbus
zur Erlangung des akademischen Grades eines
Doktor-Ingenieurs
genehmigte Dissertation

vorgelegt von

Diplom-Ingenieurin
Linda Maria Beck
aus Cottbus.

Gutachter: Univ.-Prof. Dr.-Ing. Fabian Mauß
Gutachter: Univ.-Prof. Dr.-Ing. Konstantinos Boulouchos
Gutachter: Univ.-Prof. Dr.-Ing. Heinz Peter Berg

Tag der mündlichen Prüfung: 23. Mai 2013

Bibliografische Information der Deutschen Nationalbibliothek

Die Deutsche Nationalbibliothek verzeichnet diese Publikation in der
Deutschen Nationalbibliografie; detaillierte bibliografische Daten sind
im Internet über http://dnb.d-nb.de abrufbar.

ISBN 978-3-8325-3426-4

Logos Verlag Berlin GmbH
Comeniushof, Gubener Str. 47,
10243 Berlin
Tel.: +49 (0)30 42 85 10 90
Fax: +49 (0)30 42 85 10 92
INTERNET: http://www.logos-verlag.de

Publications about the content of this work require the written consent of Volkswagen AG. The results, opinions and conclusions expressed in this thesis are not necessarily those of Volkswagen AG.

Acknowledgement

This book is the final result of a three (plus one) years work whichs root idea was already born before I found my feet in a large international corporation, rather than engine simulation, and not to mention combustion modelling issues. Some years have elapsed, and the Volkswagen Research Group gave me the opportunity to work on this thesis. For this I thank first of all Axel Winkler, Jürgen Willand, and Michael Frambourgh.

Following a promising kickoff in Prague, the time passed and the thesis progress turned more and more into a daily routine of read science, think science, do science, get error code 171, get *error because of too many errors*, get results, refuse results, and go back to start. It requires alot of knowledge, ideas, persistency, attention, and affection, to get out of this vicious circle. But thanks to the persistent help of Axel's group - a special thanks to Ema for her initial help - I learned to cope with it, re-trying again and again to finalise science.

However, the work behind was not an one-man business, and I owe a debt of gratitude to all the people, who supported this work in howsoever: Andreas, Henrik and Dr. Theobald for their interest and valuable information; the Loge team - Karin, Harry, Lars, Cathleen and Anders - for their tireless support; and the Ricardo Vectis team around Nick for their efforts trying to get everything done and dusted.

During this time, I got to know and to work with two really special individuums - Fabian and Tao. I gratefully thank both of them for the hot tempered but absolutely informative and inspiring meetings. Tao - thanks for your pragmatic thinking, outspokenness, and coding - I still cherish your goodbye-version. And Fabian - thank you for your constant support, enthusiasm, ideas, and finally for helping me to get everything to an happy ending.

But I also owe this happy ending Prof. Konstantinos Boulouchos and Prof. Heinz Peter Berg. Thank you for reviewing this thesis and the appealing finalisation.

All in all, it was a great experience but also a quite tough time, which gave me also the opportunity to get to know strange beings - the old ship, the red lion, and the green monster. Thank you for your inspiration.

Abstract

The virtual development of future Spark Ignition (SI) engine combustion processes in three-dimensional Computational Fluid Dynamics (3D-CFD) demands for the integration of detailed chemistry, enabling - additionally to the 3D-CFD modelling of flow and mixture formation - the prediction of fuel-dependent SI engine combustion in all of its complexity. The conflict of goals arising in coupling 3D-CFD calculations with detailed chemistry is to keep computational costs low while achieving accurate results.

This work presents an approach, which constitutes a coupled solution for flame propagation, auto-ignition, and emission formation modelling incorporating detailed chemistry, while exhibiting low computational costs.

For modelling the regular flame propagation, a laminar flamelet approach, the G-equation is used. This approach describes the flame propagation based on the turbulent flame speed, which is determined by the turbulence and the fuel-specific laminar flame speed. The latter one is incorporated using an adequate fitting function.

Auto-ignition phenomena are addressed using an integrated flamelet approach, which bases on the tabulation of fuel-dependent reaction kinetics. By introducing a progress variable for the auto-ignition - the Ignition Progress Variable (IPV) - detailed chemistry is integrated in 3D-CFD. The tabulation approach only demands for the solution of the IPV transport equation, thus keeping the computational demand low, while allowing the consideration of local effects on auto-ignition chemistry on cell level.

The modelling of emission formation bases on an interactively coupled flamelet approach, the Transient Interactive Flamelet (TIF) model. By transforming the species balance equations into a one-dimensional form, the numerical effort incorporated with the solution of small chemical time scales is separated from the 3D-CFD flow field solution. Thus, the emission formation is calculated under representative boundary conditions. The description of the soot formation bases on a detailed soot model, and the properties of the soot Particle Size Distribution Function (PSDF) are calculated using the method of moments.

The coupling between the G-equation, integrated flamelet, and interactive flamelet models is done based on the IPV. The functionality of the combined approach to model the variety of SI engine combustion phenomena is proved first in terms of fundamentals and standalone sub-model functionality studies. For standalone and model coupling functionality studies, a simplified test case is introduced, representing an adiabatic pressure vessel without moving meshes. The vessel is initialised homogeneously, allowing the selective investigation of different parameters on combustion process and direct comparison with direct numerical solution of the detailed chemistry in 0D homogeneous reactor calculations. Following the basic functionality studies, the standalone and combined sub-model functionalities are investigated and validated in adequate engine test cases.

For the validation of the G-equation model, a global fuel-air equivalence ratio variation of a highly turbocharged SI engine is considered. Compared with pressure trace measurements, the model matches the measured in-cylinder pressure during the combustion process with a maximum Mean Absolute Percentage Error (MAPE) of 0.02 %. Furthermore, the calculated flame fronts match optical measurements of the flames in terms of shape and local propagation speed well. The validation of the coupled G-equation and IPV model bases on in-cylinder pressure measurements of a spark-assisted Homogeneous Charge Compression Ignition (HCCI) engine. The model predicts the measured in-cylinder pressure during the combustion process with a maximum MAPE of 3.88 %. However, high standard deviations of in-cylinder pressure measured restrict the model validation. Furthermore, the functionality of the auto-ignition model to predict pre-ignition phenomena is investigated for a tumble flap variation of a highly turbocharged SI engine. The calculated pre-ignition tendencies and locations match the optical measurement data well, although the calculated total number of pre-ignition events in a given time period exceeds the measurement data. The functionality of the coupled G-equation, integrated flamelet and interactive flamelet model is finally shown for a Start Of Injection (SOI) variation of an optical SI engine. The predicted soot tendencies and origins are confirmed by optical measurements, although the relative difference of the particle numbers measured is, in comparison with the difference of the calculated particle numbers, smaller.

Kurzfassung

Für die Brennverfahrensentwicklung zukünftiger Ottomotoren mit Hilfe dreidimensionaler numerischer Strömungssimulationen (3D-CFD) ist es unabdingbar, detaillierte Reaktionskinetik in die 3D-CFD Prozesskette zu integrieren, um zusätzlich zur Modellierung der Strömung und Gemischbildung auch den kraftstoffabhängigen Verbrennungsprozess in all seiner Komplexität voraussagen zu können. Zielkonflikt bei der Kopplung von 3D-CFD Berechnungen mit Reaktionskinetik ist, bei großer Genauigkeit einen vertretbaren Rechenaufwand zu realisieren.

In dieser Arbeit wird ein Ansatz vorgestellt, welcher eine gekoppelte Lösung der Modellierung des Flammenfortschrittes, der Selbstzündung, sowie der Emissionsbildung unter Berücksichtigung der detaillierten Reaktionskinetik darstellt, und zusätzlich geringe Rechenzeiten aufweist.

Für die Modellierung der regulären Flammenausbreitung wird auf einen laminaren Flamelet Ansatz, die G-Gleichung, zurückgegriffen. Diese beschreibt die Flammenausbreitung basierend auf der turbulenten Brenngeschwindigkeit, welche durch die Turbulenz und die kraftstoffabhängige laminare Brenngeschwindigkeit beeinflusst wird. Letztere wird durch eine entsprechende Ersatzfunktion in die 3D-CFD Software integriert.

Selbstzündungsphänomene werden mit einem integrierten Flamelet Ansatz beschrieben. Dabei erfolgt die Lösung der detaillierten Reaktionskinetik vorab und die 3D-CFD Software interpoliert die vorausberechneten reaktionskinetischen Ergebnisse. Hierzu wird eine Fortschrittsvariable für die Selbstzündung eingeführt - die Ignition Progress Variable. Somit kann das Modell ressourcengünstig Selbstzündungsphänomene in der unverbrannten Mischung unter Berücksichtigung lokaler Effekte voraussagen.

Die Modellierung der Emissionsbildung basiert auf einem interaktiv gekoppelten Flamelet Ansatz, dem Transient Interactive Flamelet Modell. Durch die Transformation der Speziesbilanz- und Enthalpieerhaltungsgleichungen in eindimensionale Form wird der numerische Aufwand,

der mit der Lösung der kleinen chemischen Zeitskalen verbunden ist, von
der Lösung des 3D-CFD Strömungsfeldes getrennt. Gleichzeitig können
die Emissionen unter Berücksichtigung repräsentativer Randbedingungen berechnet werden. Für die Modellierung der Rußbildung wird ein detailliertes Rußmodell verwendet und die Eigenschaften der Rußpartikel-Größenverteilungsfunktion mit Hilfe der Momentenmethode berechnet.
Die Kopplung zwischen dem G-Gleichungs-, integriertem Flamelet-,
und interaktivem Flamelet-Modell basiert auf der Selbstzündungsvariable. Die Funktionalität des gekoppelten Ansatzes für die Modellierung
der Vielfalt an ottomotorischen Verbrennungsphänomenen, wird zuerst
in grundsätzlichen Untersuchungen unabhängig voneinander betrachtet.
Für die Untersuchung der unabhängigen als auch gekoppelten Modellfunktionalität wird ein vereinfachter Testfall eingeführt, welcher einen
adiabaten Druckbehälter ohne bewegte Gitter darstellt. Dieser Behälter
wird homogen initialisert, was eine selektive Untersuchung verschiedener
Paramter auf den Verbrennungsprozess ermöglicht. Darüber hinaus erlaubt der vereinfachte Testfall einen direkten Vergleich der 3D-CFD Lösung mit direkten numerischen Simulationen der detaillierten Chemie
in null-dimensionalen homogenen Reaktormodellen. Desweiteren wird
die Funktionalität der unabhängigen und gekoppelten Modelle anhand
entsprechender Motortestfälle untersucht und validiert.
Für die Validierung des G-Gleichungsmodells wird eine Variation
des globalen Kraftstoff-Luft-Verhältnisses an einem hoch aufgeladenen
Ottomotor betrachtet. Die berechneten Zylinderdruckverläufe weichen
dabei von den gemessenen Druckverläufen während des Verbrennungsprozesses nur geringfügig, mit einem mittleren absoluten prozentualen
Fehler von 0.02 %, ab. Desweiteren stimmen die berechneten Flammenfronten sowohl in Bezug auf Form als auch lokale Ausbreitungsgeschwindigkeit gut mit optisch gemessenen Flammenfronten überein.
Die Validierung des gekoppelten G-Gleichungs- und IPV-Modells basiert
auf einem zündfunken-gestützten homogenen Kompressionszündungs-
Brennverfahrens. Das Modell sagt dabei die gemessenen innermotorischen Druckverläufe während des Verbrennungsprozesses mit einem
mittleren absoluten prozentualen Fehler von 3.88 % voraus. Hohe Standardabweichungen der am Prüfstand gemessenen Druckverläufe beschränken jedoch die Validierungsmöglichkeit. Die Funktionalität des Selbstzündungsmodells für die Voraussage von Vorentflammungsphänomenen
wird anhand einer Variation der Ladungsbewegungsklappen-Stellung an
einem hoch aufgeladenen Ottomotor untersucht. Die berechneten Tendenzen und Vorentflammungsorte stimmen dabei gut mit den optischen
Messungen überein, auch wenn in einem gegebenen Zeitintervall die

berechnete Anzahl der Vorentflammungsereignisse die Anzahl der gemessenen Ereignisse übersteigt. Die Funktionalität des gekoppelten G-Gleichungs-, integriertem Flamelet- und interaktivem Flamelet-Modell wird schließlich anhand einer Variation des Einspritzbeginns an einem optischen Motor aufgezeigt. Die prädiktierten Rußtendenzen und -quellen stimmen gut mit optischen Messungen überein, auch wenn der relative Unterschied zwischen den gemessenen Partikelanzahlen im Vergleich mit dem berechneten Unterschied geringer ist.

Contents

Nomenclature

Greek Symbols

α	Angle	$^\circ$
β	Collision Frequency Factor	$1/(\mathrm{m^3\,s})$
χ	Scalar Dissipation Rate	$1/\mathrm{s}$
Δ	Roughness	m
δ	Inner Flame Layer Thickness	m
Γ	Bounding Surface	$\mathrm{m^2}$
γ	Surface Tension	$\mathrm{kg/s^2}$
γ_g	Gas Mean Free Path	m
λ	Thermal Conductivity	$(\mathrm{kg\,m})/(\mathrm{K\,s^3})$
μ	Dynamic Viscosity	$\mathrm{kg/(m\,s)}$
ν	Kinematic Viscosity	$\mathrm{m^2/s}$
ν_k	Stoichiometric Coefficient	-
ω	Source Term	$\mathrm{kg/(m^3\,s)}, 1/\mathrm{s}, -$
ϕ	Fuel-Air Equivalence Ratio	-
ψ	Mass fraction EGR	-
ρ	Density	$\mathrm{kg/m^3}$
σ	Variance	-
τ	Time Scale	s
ε	Turbulent Velocity of Dissipation	$\mathrm{m^2/s^3}$

Roman Symbols

A	Area	m^2
A_j	NASA Polynomial of Mixture	-
a_j	NASA Polynomial of Species	-
A_k	Pre-Exponential Factor	$mol/(l\,s), 1/s, l/(mol\,s)$
C	Constant	-
c	Progress Variable	-
c_p	Specific Heat Capacity at Constant Pressure	$m^2/(s^2\,K)$
D	Diffusion Coefficient	m^2/s
d	Diameter	m
E	Energy	$(m^2\,kg)/(s^2)$
E_a	Activation Energy	$(m^2\,kg)/(s^2\,mol)$
f_V	Soot Volume Fraction	m^3/m^3
$f_{\tilde{c}}$	Weighting Coefficient of c	-
G	Scalar describing Flame Front	-
H	Total Enthalpy	$(kg\,m^2)/s^2$
h	Specific Enthalpy	m^2/s^2
I	Identity Matrix	-
j_i	Diffusive Flux	$kg/(m\,s)$
j_q	Heat Flux	$(kg\,m)/s^3$
k	Turbulent Kinetic Energy	m^2/s^2
k_k	Reaction Rate Coefficient	$1/s$
l	Length Scale	m
l_t	Integral Length Scale	m
m	Mass	kg
M_r	Moment of Order r	$1/m^3$

M_W	Molecular Weight	kg/mol
N	Number	-
n	Normal Vector of Unity	-
p	Pressure	$kg/(m\,s^2)$
q	Heat Flux	$(kg\,m^2)/s^3$
R	Mixture Gas Constant	$(m^2\,kg)/(s^2\,mol\,K)$
s	Flame Propagation Speed	m/s
T	Temperature	K
t	Time	s
u'	Turbulent Velocity	m/s
V	Volume	m^3
v	Velocity	m/s
w_k	Reaction Rate	$mol/(m^3\,s)$
X	Mole Fraction	-
x	Position Vector	-
Y	Mass Fraction	-
Z	Mixture Fraction	-

Non-Dimensional Numbers

Da   Damköhler Number

Ka   Karlovitz Number characterising impact of turbulence on thermal flame thickness

Ka_δ Karlovitz Number characterising impact of turbulence on inner layer thickness

Kn   Knudsen Number

Le   Lewis Number

Nu   Nusselt Number

Pr   Prandtl Number

Re   Reynolds Number

Re_t Turbulent Reynolds Number

Sc   Schmidt Number

Sh   Sherwood Number

We   Weber Number

Constants

$g = 9.80655 \text{ m/s}^2$ Gravitational Acceleration

$k_B = 1.38065 \cdot 10^{-23} \text{ (kg m}^2)/(\text{s}^2\,\text{K})$ Boltzmann Constant

$N_A = 6.02214 \cdot 10^{23} \text{ 1/mol}$ Avogadro Constant

$R_{Universal} = 8.31446 \text{ (m}^2\,\text{kg})/(\text{s}^2\,\text{mol\,K})$ Universal Gas Constant

Subscripts

0	Reference
δ	Inner Flame Layer
b	Burnt
boi	Boiling
c	Chemical
conv	Convection
cr	Critical
d	Droplet
dep	Deposition
diff	Diffusion
f	Flame
F	Fuel
g	Gas
i	Related to Species i
int	Interface
k	Kolmogorov Scale
kern	Kernel
L	Leidenfrost
l	Laminar
m	Mixture
Ox	Oxidizer
p	Particle
r	Relative
rad	Radiation

ran	Random
s	Soot
sec	Secondary
st	Stoichiometric
surf	Surface
t	Turbulent
th	Thermal
tot	Total
u	Unburnt
v	Vapour
w	Wall
Wf	Wall Film

Abbrevations

0D	0-Dimensional related to space coordinates
3D	3-Dimensional related to space coordinates
aTDC	after TDC
bTDC	before TDC
BDC	Bottom Dead Center
BMEP	Break Mean Effective Pressure
BML	Bray-Moss-Libby model
CD	Cheng-Diringer model
CFD	Computational Fluid Dynamics
CFM	Coherent Flame Model
CH	Choi-Huh model
CHP	Combined Heat and Power
CN	Cetane Number
CPB	Cant-Pope-Bray model
DDM	Discrete Droplet Model
DNS	Direct Numerical Simulation
EBU	Eddy Break-Up model
ECMF	Extended Coherent Flame Model
ECMF-3Z	3-Zone Extended Coherent Flame Model
EDM	Eddy Dissipation Model
EGR	Exhaust Gas Recirculation
ETAB	Enhanced Taylor Analogy Breakup model
FMEP	Frictional Mean Effective Pressure
FPV	Flamelet Progress Variable model

GCI	Gasoline Compression Ignition engine
HACA	Hydrocarbon-Abstraction C_2H_2-Addition
HACARC	Hydrocarbon-Abstraction C_2H_2-Addition Ring-Closure
HC	unburnt Hydro-Carbon
HCCI	Homogeneous Charge Compression Ignition
HT	High Temperature reaction pathway
IMEP	Indicated Mean Effective Pressure
IPV	Ignition Progress Variable
ISAT	In-Situ Adaptive-Tabulation model
KH-RT	Kelvin Helmholtz-Rayleigh Taylor model
LES	Large Eddy Simulation
LT	Low Temperature reaction pathway
MAPE	Mean Absolute Percentage Error
MB	Mantel-Borghi model
NO_x	Nitrogen Oxides
NTC	Negative Temperature Coefficient
OP	Operating Point
PAH	Polycyclic Aromatic Hydrocarbon
PDF	Probability Density Function
PSDF	Particle Size Distribution Function
RANS	Reynolds-Averaged Navier-Stokes equations
RON	Research Octane Number
RPM	Rotation Per Minute
SDF	Size Distribution Function
SI	Spark Ignition

SOI	Start Of Injection
TAB	Taylor Analogy Breakup model
TDC	Top Dead Center
TIF	Transient Interactive Flamelet
TKI	Tabulated Kinetics of Ignition model

Chapter 1

Introduction

1.1 Background

Due to the increasing shortage of resources the demand for energy efficiency and sustainable mobility needs to be considered continuously. Thereby, electrical drivetrains and optimised internal combustion engines as well as their operation with alternative fuels are focused [63]. In order to compare the different approaches, the whole process chain needs to be examined holistically, from the well of production to the usage of energy in the vehicle. These studies are called well-to-wheel analysis.

In principle, electrical driven vehicles do not emit CO_2 during usage. The overall CO_2 emission values depend on the production of electricity supplied and its distribution. Figure 1.1 illustrates the CO_2 emissions per kilometre of an electrical driven vehicle based on different primary energy sources used for electricity production.

The CO_2 emissions can be reduced strongly using electricity based on regenerative energy. According to [63], nowadays the so-called green electricity contributes less than 15 % to the overall energy source mix of electricity. Some optimistical prognosis indicate an increase up to 25 % in 2020. Based on the energy source mix of electricity present in Europe and its associated emissions, the well-to-wheel analysis possesses CO_2 emissions of an electrical driven vehicle of 89 g/km. This value can potentially be decreased to 58 g/km in 2020. Based on the energy source mix of electricity present in China, the same vehicle emits 179 g/km. Former value is only marginally lower than the CO_2 emission value of a diesel powered equivalent.

Well-to-wheel analysis of alternative fuels strongly depend on process energy supplied during fuel production, which becomes apparent from figure 1.2.

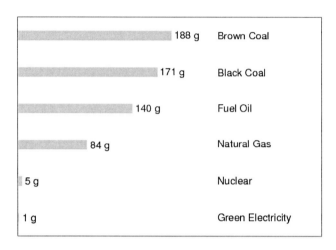

Figure 1.1: Well-to-wheel CO_2 emissions per kilometre of electric vehicles according to [63]

In the best case, this energy bases on regenerative resources increasing the CO_2 reduction potential of the so-called biofuels. For instance, wheat extracted ethanol possesses a CO_2 reduction potential of 33 % using process energy supplied by natural gas. Whereas, using process energy supplied by straw, the CO_2 reduction potential increases up to 69 %.

Due to an increasing human population (and associated food competition) and environmental requirements, the substitution potential of biofuels will decrease in future terms. Simultaneously, the individual mobility will increase world wide. Hence, regenerative fuels will not be the sole long-term solution in terms of sustainable mobility, and the importance of electrical driven vehicles will increase strongly. However, the short and mid-term substitution potential of biofuels is comparable to the one of electrical driven vehicles [63]. As a consequence, in short-term future the market share of regenerative fuels will increase and optimised internal combustion engines need to be investigated in respect to fuel dependency.

An optimised internal combustion engine is characterised by strongly reduced air pollutants, as well as low fuel consumption and low associated CO_2 emissions. Both, emission formation and fuel consumption are determined by fuel type and composition. For modern Spark Ignition (SI) engines, the downsizing strategy with direct injection is predominant [186]. The accompanying enhancement of the mean effective pressure at low revolutions and high loads is limited by stochastic auto-ignition phenomena, called pre-ignition or super knock, which are in turn sensitive to fuel composition[1].

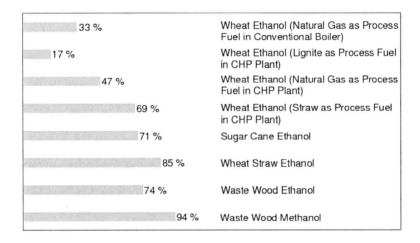

33 %	Wheat Ethanol (Natural Gas as Process Fuel in Conventional Boiler)
17 %	Wheat Ethanol (Lignite as Process Fuel in CHP Plant)
47 %	Wheat Ethanol (Natural Gas as Process Fuel in CHP Plant)
69 %	Wheat Ethanol (Straw as Process Fuel in CHP Plant)
71 %	Sugar Cane Ethanol
85 %	Wheat Straw Ethanol
74 %	Waste Wood Ethanol
94 %	Waste Wood Methanol

Figure 1.2: CO_2 reduction potential of biofuels according to [173]

In the virtual development of future combustion processes three-dimensional Computational Fluid Dynamics (3D-CFD) are an important tool. By visualising local appearing effects potentials of different combustion strategies can be evaluated. In order to account for future engine requirements, the fuel specific chemistry needs to be integrated in the 3D-CFD work flow. However, the complexity and diversity of desired and undesired SI engine combustion phenomena requires a comprehensive method development.

---

[1]For instance, the enhancement of the mean effective pressure is sensitive to ethanol blends [84].

1.2   Thesis Objectives

The objective of this thesis is to provide a solution which allows - additionally to the 3D-CFD modelling of flow and mixture formation - the prediction of fuel-dependent SI engine combustion in all of its complexity.

The SI engine combustion processes can be classified in regular combustion and irregular combustion. In case of regular (conventional) combustion, the mixture in the combustion chamber is transformed by a propagating flame front initiated by the spark plug. Irregular combustion is a result of auto-ignition phenomena in the unburnt mixture and is in case of conventional operating engines strictly undesired, since it can cause serious engine damage [70]. However, some combustion processes designate auto-ignition processes in the unburnt mixture, like Homogeneus Charge Compression Ignition (HCCI) combustion. Both combustion pathways lead to a transformation of the fuel-air mixture determined by the reaction progress and thus to an emission formation in the burning and burnt zone.

In order to simulate the fuel-dependent regular SI engine combustion, the combustion model requires a method to model the propagation of the flame initialised by the spark plug[1]. The flame propagation model needs to account for the influence of turbulence, fuel-air equivalence ratio, thermodynamics, and fuel composition. Moreover, to model auto-ignition phenomena, like HCCI combustion, knock and pre-ignition, the model needs to account for precursor reactions, thermodynamic condition and fuel composition. The modelling of emission formation includes chemical reactions as well as physical processes.

The aim of this work is to exploit the potential of a combined G-equation / integrated flamelet / interactive flamelet modelling approach in order to model the complexity and diversity of desired and undesired fuel specific SI engine combustion processes. The functional attribution of the sub-models is illustrated in figure 1.3.

For modeling the regular flame propagation, a laminar flamelet approach, the G-equation of Peters [134] is used. The model has been derived on purpose to simulate regular flame propagation in SI engines and become generally accepted. Auto-ignition phenomena are addressed using an integrated flamelet approach, the Ignition Progress Variable (IPV) model. The model was first published in [99] and applied to

---

[1]In the modelling approach applied in this work, the combustion process starts with an initialised flame kernel.

model diesel engine ignition processes. The modelling of emission formation bases on an interactive flamelet approach, the Transient Interactive Flamelet (TIF) model [121], which is coupled with a detailed soot model. The model has been applied to simulate auto-ignition as well as emission formation in diesel engine applications.

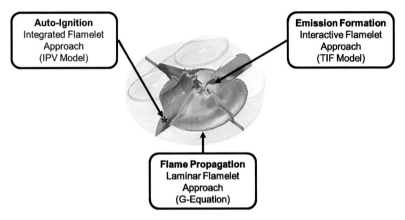

Figure 1.3: Coupled G-equation, integrated flamelet (IPV) and interactive flamelet (TIF) modelling approach

For the purpose of modelling the fuel-dependent SI engine combustion process, the diesel engine specific models need to be adjusted. Moreover, an adequate interface needs to be elaborated, enabling the modelling of flame propagation, auto-ignition as well as emission formation simultaneously. Furthermore, the model functionality to model the variety of regular and irregular SI engine combustion phenomena needs to elaborated in adequate engine test cases.

1.3   Methodology

In order to prove the functionality of the combined G-equation / integrated flamelet / interactive flamelet approach to model the variety of SI engine combustion phenomena, the fundamentals and standalone submodel functionalities are investigated first. Moreover, the interaction between the sub-models needs to be elaborated. For standalone and model coupling functionality studies, a simplified test case is introduced, representing an adiabatic pressure vessel without moving mesh. The simplified test case is displayed in figure 1.4. Details about geometrical dimensions can be found in appendix A.1.2.

0   ϕ [-]   1

Figure 1.4: Developing flame fronts (blue iso-surfaces) and scalar plane of fuel-air equivalence ratio in homogeneously initialised simplified test case

The vessel is initialised homogeneously, allowing the selective investigation of different parameters on combustion process and direct comparison with 0D homogeneous reactor calculations. Following the basic functionality studies, the standalone and combined sub-model functionalities are investigated in engine test cases.

All investigations base on an *iso*-Octane / *n*-Heptane reaction mechanism of Ahmed et al. [6], which consists of 306 chemical species.

1.4   Outline

Following a short introduction to 3D-CFD flow and mixture formation modelling fundamentals, the combustion sub-models are addressed successively.

First, the flame propagation modelling and its implementation in 3D-CFD are presented. The combined G-equation / integrated flamelet / interactive flamelet approach underlies the flamelet assumption. For this reason, the validity of this assumption for turbocharged direct injection SI engines is elaborated. The G-equation modelling approach demands for closure formulations for the laminar and turbulent flame speed. Different formulations are examined and investigated in engine test case. Afterwards, the G-equation model is validated using optical measurements.

In the next chapter, the auto-ignition modelling approach and the coupling to 3D-CFD code are addressed. The IPV model bases on the tabulation of fuel-dependent reaction kinetics. The results obtained applying tabulation methods depend strongly on the accuracy of tabulation. For this reason, the library accuracy is investigated. In order to model engine knock and HCCI combustion, the IPV model needs to be coupled with the flame propagation model. The functionality of the coupled G-equation / IPV model is investigated in terms of a HCCI combustion. Furthermore, the model capability to predict pre-ignition phenomena is investigated in a highly turbocharged SI engine, which is feasible especially for the investigation of this stochastic auto-ignition phenomena.

Afterwards, the emission formation model and its coupling to 3D-CFD is shown. To model the emission formation in burnt zone regions, the interactive flamelet model is coupled to the G-equation / IPV model. This procedure demands for a burnt flamelet initialisation, which needs to be elaborated. The model needs to account for gas-phase mixture inhomogeneities and evaporating wall film, increasing the soot formation process. Therefore, the functionality of the interactive flamelet model to predict emission formation is investigated for a variation of Start Of Injection (SOI) possessing different wall film formation tendencies in an optical SI engine.

In the last chapter, conclusions are finally drawn and collected in a summary.

Chapter 2

Fundamentals

The basis of 3D-CFD modelling of in-cylinder combustion processes constitutes the simulation of the intake and in-cylinder gas flow, and the mixture formation process, including liquid fuel injection, fuel droplet breakup, and droplet evaporation. The description of the unsteady, compressible, chemical reactive, turbulent, three-dimensional two-phase flow with moving meshes bases on conservation equations of the gaseous and the liquid phase. After a brief characterisation of the turbulent flow field in section 2.1, the underlying equations and sub-models are outlined in sections 2.2 and 2.3. For a detailed description see [77, 118, 132, 154].

2.1 Phenomenological Description of Turbulent Flow

The in-cylinder engine combustion process takes place in a turbulent flow field [70, 136]. In contrast to laminar flow fields where fluid flows in parallel layers with no disruption between the layers, turbulent flows are unsteady, three-dimensional, vortical, and random [136].

A laminar flow becomes turbulent when destabilising inertial forces dominate stabilising viscosity forces, or in other words, the convective process τ_{conv} is much faster than the molecular diffusion process τ_{diff} [70, 77]. The ratio of both processes is related in the non-dimensional Reynolds number Re.

$$\frac{\tau_{diff}}{\tau_{conv}} \sim \mathrm{Re} = \frac{v\, l}{\nu} \tag{2.1}$$

Apparent from the Reynolds number definition, the transition from laminar to turbulent state occurs with increasing solid object size l, increasing flow velocity v, or lower fluid kinematic viscosity ν [136]. In common the increased convection process is a result of shear in the mean flow [70]. The shear leads to the formation of eddies causing turbulent velocity fluctuations u', which in turn lead to fluctuations of scalar variables, like density, temperature, and concentration [77].

Turbulent flows are characterised by a broad spectrum of large-scale (high wavelength, low frequency) and small-scale (low wavelength, high frequency) eddy structures [53]. Disruptions of high wavelength primarily lead to the formation of large-scale eddies. The largest scales correspond to the geometrical dimension [77] and are described by the integral length scale l_t. The large-scale eddy structures directly interact with the mean flow and absorb turbulent kinetic energy k. The eddies interact with each other and decay under the formation of progressively smaller eddies [77]. Thereby, the large-scale eddies supply in a non-linear energy transfer turbulent kinetic energy to the small-scale eddies. However, the turbulent kinetic energy concentrates in large-scale eddies and decreases strongly with decreasing wavelength [69]. This transformation process is also known as cascade theory of Kolmogorov [90].

The wavelength spectrum is limited due to an increasing impact of the molecular viscosity with decreasing wavelength, leading to an increasing dissipation ε of the turbulent kinetic energy into thermal energy. The smallest eddy structures sustaining in a turbulent flow field are of Kolmogorov length l_k. Their size is dependent on the viscosity of the fluid and the specific turbulent dissipation [69].

$$l_k = \left(\frac{\nu^3}{\varepsilon}\right)^{1/4} \tag{2.2}$$

Following Kolmogorov's theory, which assumes an equilibrium of transferred and dissipated energy, the dissipation of the turbulent kinetic energy is dominated by the large-scale motions [67], i.e.

$$\varepsilon = \frac{u'^3}{l_t} \tag{2.3}$$

with

$$u' = \left(\frac{2k}{3}\right)^{1/2} \tag{2.4}$$

although the dissipation is a result of viscosity forces at small scales.

For a more adequate description of the intensity of turbulence Williams [190] introduced the turbulent Reynolds number

$$\mathrm{Re}_t = \frac{u' l_t}{\nu} \tag{2.5}$$

which relates the integral length scale and the Kolmogorov length scale.

$$\mathrm{Re}_t^{3/4} = \frac{l_t}{l_k} \qquad (2.6)$$

Both, large-scale and small-scale turbulence motions determine the overall behaviour of a turbulent flow field [70]. Simulation methods for the calculation of the turbulent flows are the Direct Numerical Simulation (DNS), Large Eddy Simulation (LES) and Reynolds-Averaged Navier-Stokes (RANS) equations [53]. The methods distinguish in terms of level of detail capturing the different turbulent scales.

In DNS all length scales up to the smallest structures of the turbulent flow field are calculated. The direct calculation of the flow field up to the smallest length scales demands for a fine discretisation in time and space. Due to the high computational costs involved, the application of this approach is nowadays still limited to simplified geometries and low Reynolds numbers [178].

Large-scale eddy structures are difficult to model due to their non-isotropy. Small eddy structures are generated by dissipation of the large eddy structures and are thus isotropic, enabling the modelling of these structures. This advantage is pursued in LES, by explicitly calculating the large-scale eddy structures and modelling the small-scale structures. By modelling the small-scale structures, LES demands in comparison to DNS for less mesh refinement [53]. Nevertheless, for in-cylinder engine applications the computational demand is nowadays still to high.

Oftentimes, averaged values of the in-cylinder engine process are requested [77]. Due to the irregularity and randomness of turbulent flows, an average of the in-cylinder engine process can be obtained using DNS or LES only by a large number of calculations under marginally varying initial and boundary conditions [53]. For the industrial application, a statistical method, the RANS approach, is therefore used to define such a flow field [70].

2.2   Gas Phase Modelling

The RANS approach considers the turbulence statistically on the basis of Reynolds-averaged Navier-Stokes equations [53]. All time fluctuations are removed by an averaging procedure. Thus, only the statistical equations of the average need to be solved. The fluctuations appear in the equations as average terms, which need to be modelled by a turbulence model.

2.2.1 Averaging of Flow Variables

To account for the turbulence impact it is convenient to introduce statistically averaged flow variables into the conservation equations. For the statistical description of the turbulence of unsteady flows, ensemble averaging is applied. An ensemble average $\overline{\Phi}$ of a variable Φ describes the arithmetic mean of a multitude of measurements N constituting identical initial and boundary conditions [132].

$$\overline{\Phi}\left(\mathbf{x},t\right) = \lim_{N \to \infty} \frac{1}{N} \sum_{i=1}^{N} \Phi_i\left(\mathbf{x},t\right) \qquad (2.7)$$

For turbulent flows possessing large density change it is convenient to introduce a density-weighted (Favre) averaging. Whereat, the statistical average of a flow variable Φ is expressed by its mean value $\widetilde{\Phi}$ and fluctuating part $\Phi^{''}$ [132].

$$\Phi\left(\mathbf{x},t\right) = \widetilde{\Phi}\left(\mathbf{x},t\right) + \Phi^{''}\left(\mathbf{x},t\right) \qquad \text{with} \qquad \overline{\rho \Phi^{''}} = 0 \qquad (2.8)$$

The resulting Favre-averaged variables are introduced in the conservation equations of the gas phase.

2.2.2 Gas Phase Conservation Equations

The description of the continuum flow of a compressible Newtonian fluid bases on the Navier-Stokes equations of the extensive variables total mass, momentum, species mass, and energy. The Navier-Stokes equations are presented in Favre-averaged form in the following.

Continuity

The continuity equation describes the conservation of the total mass in a system and reads in Favre-averaged form

$$\frac{\partial}{\partial t}\overline{\rho} + \nabla \cdot \left(\overline{\rho}\,\widetilde{\mathbf{v}}\right) = \overline{\rho}\,\widetilde{\omega}_{spray}. \qquad (2.9)$$

Herein, the mass exchange between the liquid and the gas phase due to droplet evaporation and vapour condensation is described by the source term $\widetilde{\omega}_{Spray}$.

Momentum

The momentum equation reads

$$
\frac{\partial}{\partial t}\left(\overline{\rho}\,\widetilde{\mathbf{v}}\right) + \nabla\cdot\left(\overline{\rho}\,\widetilde{\mathbf{v}}\,\widetilde{\mathbf{v}}\right) = -\nabla\overline{\mathbf{p}} + \nabla\overline{\tau} - \nabla\cdot\left(\overline{\rho}\,\widetilde{\mathbf{v}''\mathbf{v}''}\right) + \overline{\rho}\,\mathbf{g} + \widetilde{\omega}_{Spray}. \quad (2.10)
$$

The first term on the left hand side represents the locale change rate, the second term the convection of momentum [132]. The change rate of momentum is determined by the pressure gradient $\overline{\mathbf{p}}$, the molecular transport due to viscosity $\overline{\tau}$, forces due to buoyancy $\overline{\rho}\,\mathbf{g}$, and momentum gained per unit volume due to the interaction with the liquid phase $\widetilde{\omega}_{Spray}$.

The Favre-averaging results in the momentum equation in an additional term $\overline{\rho}\,\widetilde{\mathbf{v}''\mathbf{v}''}$, the so-called Reynolds stress tensor, describing the convective momentum transport by turbulent fluctuations. The Reynolds stress tensor constitutes the first closure problem for turbulence modelling.

Reactive Scalar

Considering a mixture of N chemical reacting species, the balance equation of mass fraction of species i ($1 \leq i \leq N$) reads

$$
\frac{\partial}{\partial t}\left(\overline{\rho}\,\widetilde{Y}_i\right) + \nabla\cdot\left(\overline{\rho}\,\widetilde{\mathbf{v}}\,\widetilde{Y}_i\right) = \overline{\rho}\,D_i\cdot\nabla\widetilde{Y}_i + \omega_i + \omega_{spray}. \quad (2.11)
$$

The two terms on the left hand side represent the local change rate and the convection of mass fraction of species i. The first term on the right hand side describes the diffusive flux, approximated as binary flux ($\overline{\rho}\,D_i\cdot\nabla\widetilde{Y}_i$), with D_i as the diffusion coefficient of species i. For simplification the mass diffusivity for all species is assumed to be proportional to the conductivity of temperature [132]

$$
D = \frac{\lambda}{\rho\,c_p} \quad (2.12)
$$

with λ as thermal conductivity. This assumption leads to a constant Lewis number.

$$
\mathrm{Le}_i = \frac{\lambda}{\rho\,c_p\,D_i} = \frac{D}{D_i} \quad (2.13)
$$

The last two terms in equation 2.11 denote source terms of chemical reactions and droplet evaporation.

Energy

The energy conservation is formulated in terms of the enthalpy h.

$$\frac{\partial}{\partial t}\left(\overline{\rho}\,\widetilde{h}\right) + \nabla \cdot \left(\overline{\rho}\,\widetilde{\mathbf{v}}\,\widetilde{h}\right) = -\nabla\overline{\mathbf{j_q}} - \nabla \cdot \left(\overline{\rho\,\widetilde{\mathbf{v''}h''}}\right) - \frac{\partial \overline{p}}{\partial t} + \widetilde{q}_w - \widetilde{q}_{rad} + \widetilde{q}_{spray}$$
(2.14)

The terms on the left hand side represent the local change rate and convection of the enthalpy. On the right hand side $\overline{\mathbf{j_q}}$ denotes the heat flux accounting for thermal diffusion and enthalpy transport by species diffusion $\mathbf{j}_i$.

$$\mathbf{j_q} = -\lambda \cdot \nabla T + \sum_{i=1}^{N} h_i \mathbf{j}_i$$
(2.15)

The term $\overline{\rho\,\widetilde{\mathbf{v''}h''}}$ describes the transport of convective enthalpy (turbulent scalar fluxes), which needs to be modelled as unclosed, $\widetilde{q}_w$ the Favre-averaged heat flux through the wall, $\widetilde{q}_{rad}$ the contribution of radiation, and $\widetilde{q}_{spray}$ the energy contribution of the liquid phase due to droplet evaporation.

2.2.3   Closure Formulation

The Favre-averaging of the Navier-Stokes equations results in additional terms in the momentum and the enthalpy equations, the Reynolds stresses and the turbulent scalar fluxes, physically representing the diffusive transport of momentum and enthalpy due to turbulence. To close the equation system, a turbulence model is required providing a set of approximation equations.

One of the oldest approaches of the classical turbulence theory base on the Boussinesq's [23] concept of eddy viscosity. To close the equation system, Boussinesq proposed to relate the turbulence stresses to the mean flow with a proportionality factor ν_t, the turbulent (eddy) viscosity. Introducing the turbulent viscosity, the Reynolds stresses are treated in analogy to stresses induced by molecular viscosity in laminar flows.

$$-\widetilde{\mathbf{v''}\mathbf{v''}} = \nu_t \left(\nabla\widetilde{\mathbf{v}} + \nabla\widetilde{\mathbf{v}}^T - \frac{2}{3} \cdot \nabla\widetilde{\mathbf{v}}\, I\right) - \frac{2}{3}\widetilde{k}\, I$$
(2.16)

Herein, I denotes the identity matrix and $\widetilde{k}$ the mean turbulent kinetic energy.

$$\widetilde{k} = \frac{1}{2}\widetilde{\mathbf{v''}\mathbf{v''}}$$
(2.17)

The turbulent transport term in the enthalpy equation can be closed using a gradient flux approximation.

$$-\widetilde{\mathbf{v}''h''} = -\frac{\nu_t}{\mathrm{Pr}} \cdot \nabla\widetilde{h} \tag{2.18}$$

Thus, the turbulent viscosity is taken as an isotropic property of the flow which changes in time and position. Due to dimensional reason, the turbulent viscosity can be expressed as product of turbulent velocity scale u' and length scale l_t that is characteristic of the turbulent fluctuations.

$$\nu_t = \frac{\mu_t}{\rho} \propto u' l_t \tag{2.19}$$

Determining the turbulent velocity and length scale, the Reynolds-Averaged Navier-Stokes equation system can be closed.

The different eddy viscosity based models distinguish in terms of the number of independent turbulence variables used to determine u' and l_t. In the standard $k - \varepsilon$ model, the turbulent velocity and length scales are given in a combination of turbulent kinetic energy k and its isotropic dissipation ε.

$$l_t = \frac{\widetilde{k}^{3/2}}{\widetilde{\varepsilon}} \tag{2.20}$$

$$u' = \frac{2}{3}\widetilde{k}^{1/2} \tag{2.21}$$

The turbulent viscosity reads thus

$$\nu_t = C_\mu \frac{\widetilde{k}^2}{\widetilde{\varepsilon}} \tag{2.22}$$

where C_μ is a flow dependent modelling constant assumed to be 0.09. The variables k and ε are determined by solving additional transport equations.

$$\frac{\partial}{\partial t}\left(\overline{\rho}\,\widetilde{k}\right) + \nabla \cdot \left(\overline{\rho}\,\widetilde{\mathbf{v}}\,\widetilde{k}\right) = \nabla \cdot \left[\left(\frac{\overline{\rho}\,\nu_t}{C_{\sigma k}}\right) \cdot \nabla\widetilde{k}\right] \tag{2.23}$$

$$- \left(\frac{2}{3}\overline{\rho}\,\widetilde{k} \cdot \nabla\widetilde{\mathbf{v}} + \tau : \nabla\widetilde{\mathbf{v}}\right) - \overline{\rho}\,\widetilde{\mathbf{v}''\mathbf{v}''} : \nabla\widetilde{\mathbf{v}} - \overline{\rho}\,\widetilde{\varepsilon}$$

$$\frac{\partial}{\partial t}\left(\overline{\rho}\,\widetilde{\varepsilon}\right) + \nabla \cdot \left(\overline{\rho}\,\widetilde{\mathbf{v}}\,\widetilde{\varepsilon}\right) = \nabla \cdot \left(\frac{\overline{\rho}\,\nu_t}{C_{\sigma \varepsilon}} \cdot \nabla\widetilde{\varepsilon}\right) \tag{2.24}$$

$$+ \frac{\widetilde{\varepsilon}}{\widetilde{k}}\left(C_{\varepsilon 3}\,\overline{\rho}\,\widetilde{\varepsilon} \cdot \nabla\widetilde{\mathbf{v}} + C_{\varepsilon 1}\,\overline{\rho}\,\widetilde{\mathbf{v}''\mathbf{v}''} : \nabla\widetilde{\mathbf{v}} - C_{\varepsilon 2}\,\overline{\rho}\,\widetilde{\varepsilon}\right)$$

The terms on the left hand side represent the local change rate and convection of the turbulent kinetic energy and its dissipation. On the right hand side, the first term denotes the diffusion, the second term the fluctuation, the third term the production and the last term the dissipation of k and ε, respectively. In the standard $k - \varepsilon$ model the model constants are assumed to $C_{\varepsilon 1} = 1.44$, $C_{\varepsilon 2} = 1.92$, $C_{\varepsilon 3} = 0.373$, $C_{\sigma k} = 1.0$, and $C_{\sigma \varepsilon} = 1.22$.

2.3   Liquid Phase Modelling

In direct injection engine spray simulations, the coupling between the gas phase and the liquid phase bases on a separated flow modelling approach, which considers the exchange of mass, momentum, and energy between both phases. Due to the large number of droplets a direct injection spray consists of, direct numerical calculation of the single droplets is not feasible due to high computational demand[1]. For this reason, the droplet dynamic is described by a statistical approach by introducing a Probability Density Function (PDF) [189]

$$\Psi\left(\mathbf{x}, \mathbf{v}_d, d_d, T_d, t\right) = \frac{\mathrm{d}^8 N}{\mathrm{d}\mathbf{x}\,\mathrm{d}\mathbf{v}_d\,\mathrm{d}d_d\,\mathrm{d}T_d} \qquad (2.25)$$

describing the probability of occurrence of N droplets in volume element $\mathrm{d}\mathbf{x}$, located at position $\mathbf{x}$ at time t, which are characterised by velocity $\mathbf{v}_d, \mathbf{v}_d + \mathrm{d}\mathbf{v}_d$, diameter $d_d, d_d + \mathrm{d}d_d$ and temperature $T_d, T_d + \mathrm{d}T_d$. The temporal and spatial evolution of the PDF Ψ is described by the spray equation [189].

$$\frac{\partial}{\partial t}\left(\Psi\right) + \nabla_{\mathbf{x}} \cdot \left(\Psi\, \dot{\mathbf{x}}_d\right) + \nabla_{\mathbf{v}} \cdot \left(\Psi\, \dot{\mathbf{v}}_d\right) \qquad (2.26)$$

$$+ \frac{\partial}{\partial d_d}\left(\Psi\, \dot{d}_d\right) + \frac{\partial}{\partial T_d}\left(\Psi\, \dot{T}_d\right) = \dot{\omega}_{Spray}$$

The temporal evolution terms (location $\dot{\mathbf{x}}_d$, velocity $\dot{\mathbf{v}}_d$, diameter $\dot{d}_d$, and temperature $\dot{T}_d$) represent the continuous processes[2]. The source term $\dot{\omega}_{Spray}$ describes all discontinuous processes inducing a change of droplet number, like droplet formation during spray injection, and breakup.

---

[1]Moreover, unknown initial and boundary conditions complicate direct numerical calculation.

[2]The continuous process of turbulent dispersion is not represented by the temporal evolution terms. It is rather explicitly given by the instantaneous velocity in droplet acceleration $\dot{\mathbf{v}}_d$.

The spray equation constitutes a high-dimensional integro-differential equation. An analytical solution of the spray equation thus demands for strongly restrictive assumptions. For the numerical solution two frequently used approaches are the two-fluid model and the Discrete Droplet Model (DDM). In both cases, the continuous phase is calculated with the Eulerian method. However, the approaches differ in treatment of the dispersed phase.

The two-fluid model treats the gas and the liquid phase as interacting and interpenetrating continua. Using Eulerian formulation for the dispersed phase, the PDF Ψ needs to be discretised in all dimensions. Due to enhanced computational costs[1] involved, this approach is not feasible for engine spray calculations.

In the DDM approach the fuel spray is treated as a dispersed liquid phase, which moves in and interacts with the surrounding continuous gas phase. The PDF Ψ is discretised by a finite number of representative parcels [38] consisting of a certain number of droplets having the same physical properties (size, velocity, temperature). The motion of the parcels is tracked in Lagrangian fashion as they move through the gas phase, exchanging mass, momentum and energy. The effect of the droplet parcels on the continuous phase due to drag, heat and mass transfer is accounted for via source terms in the gas phase conservation equations. To close the spray equation, the undefined terms $(\mathbf{v}_d, d_d, T_d, \dot{\omega}_{Spray})$ need to be modelled with consideration of the gas flow.

Sub-processes that need to be modelled in practical sprays typically include the liquid atomisation, droplet breakup, droplet evaporation, droplet-turbulence interaction, and spray wall impingement. Additionally, in case spray wall impingement results in deposition of the liquid fuel on the cylinder wall, a wall film model is required. The approach taken in this thesis prescribes the initial mean droplet properties and models the droplet (secondary) breakup process due to instabilities based on density- and velocity gradients between liquid and gaseous phase. The underlying equations are outlined in the following.

---

[1]Increased computational costs are a result of the high dimension and numerical diffusive character of Ψ. For the reduction a fine discretisation is required.

2.3.1   Equation of Motion

The trajectories of spherical spray droplets, moving with velocity $\mathbf{v}_d$ in a flow field, are governed by Newton's second law by balancing outer forces and inertia

$$m_d \frac{d\mathbf{v}_d}{dt} = \frac{1}{2} C_d \rho A_d \, |\mathbf{v}_r| \, \mathbf{v}_r \tag{2.27}$$

where $\mathbf{v}_r$ represents the specific force of acceleration resulting from relative movement $\mathbf{v}_r = \mathbf{v}_g - \mathbf{v}_d$ between gas and droplet, ρ is the gas density, A_d is the projected area of droplet in moving direction, and C_d the droplet drag coefficient. Assuming a rigid spherical, the droplet drag coefficient can be expressed as [143]

$$C_d = \begin{cases} 24/\mathrm{Re}_d \left(1 + 1/6 \cdot \mathrm{Re}_d^{2/3}\right) & \text{if } \mathrm{Re}_d \leq 1000 \\[2mm] 0.424 & \text{if } \mathrm{Re}_d > 1000 \end{cases} \tag{2.28}$$

where Re_d is the droplet Reynolds number

$$\mathrm{Re}_d = \frac{|\mathbf{v}_r| \, d_d}{\nu}. \tag{2.29}$$

When the Reynolds number tends to zero, the formula approaches the Stokes Law, i.e. $C_d = 24/\mathrm{Re}_d$. For high Reynolds numbers, the drag coefficient is set to a constant value.

2.3.2   Droplet Evaporation

The evaporation of droplets in a spray involves simultaneous heat and mass transfer processes. Heat that causes evaporation is transferred to the droplet surface by convection and conduction from the surrounding gas, and the vapour is transferred back to the gas phase by convection and diffusion. The droplet mass [22] and temperature are calculated through the correlations [33, 163]

$$\frac{dm_d}{dt} = -A_d \, \mathrm{Sh} \, \frac{D_{AB}}{d_d} \rho_v \, \ln \frac{(p - p_{v,\infty})}{(p - p_{v,s})} \tag{2.30}$$

$$m_d \frac{dc_{p,v} T_d}{dt} = -A_d \, \mathrm{Nu}(T_d - T) \, \lambda_m \, F_{Cz} + h_{fg} \frac{dm_d}{dt} \tag{2.31}$$

where D_{AB} is the mass diffusivity, λ_m is the mixture thermal conductivity, $c_{p,v}$ is the specific heat of liquid fuel, h_{fg} is the latent heat of evaporation, p is the pressure of gas mixture, $p_{v,s}$ and $p_{v,\infty}$ stand for the partial pressure at the droplet surface and ambient, respectively. The fuel vapour density ρ_v is evaluated from the mean temperature $\overline{T} = (T_d + T)/2$ and the gas pressure. The Sherwood Sh and Nusselt Nu numbers are calculated through the correlation by Ranz and Marshall [144]

$$\text{Sh} = 2 \cdot \left(1 + 0.3 \cdot \text{Re}_d^{1/2} \cdot \text{Sc}^{1/3}\right) \tag{2.32}$$

$$\text{Nu} = 2 \cdot \left(1 + 0.3 \cdot \text{Re}_d^{1/2} \cdot \text{Pr}^{1/3}\right) \tag{2.33}$$

where Sc and Pr are the Schmidt and Prandtl numbers.

$$\text{Sc} = \frac{\mu}{\rho \, d_d} \tag{2.34}$$

$$\text{Pr} = \frac{c_p \, \mu}{\lambda_m} \tag{2.35}$$

The quantity F_{C_Z} is defined as

$$F_{C_Z} = \begin{cases} C_Z / \left(e^{C_Z} - 1\right) & \text{if } C_Z > 0.001 \\ \\ 1.0 & \text{if } C_Z \leq 0.001 \end{cases} \tag{2.36}$$

with

$$C_Z = \frac{c_{p,v}}{\pi \, d_d \, \lambda_m} \text{Nu} \frac{\mathrm{d}m_d}{\mathrm{d}t}. \tag{2.37}$$

2.3.3   Liquid Atomisation and Droplet Size Distribution

The atomisation of in-cylinder fuel sprays can be divided into two main processes, the primary and the secondary breakup. The primary breakup takes place in regions close to the nozzle and involves liquid atomisation, and droplet breakup mechanism resulting from internal nozzle phenomena, like turbulence and cavitation. The proceeding atomisation of the primary breakup droplets occurring further downstream the spray is called secondary breakup. The secondary breakup process is largely independent of the nozzle type and determined by aerodynamic interaction processes.

The modelling of the primary breakup demands for an integrated physical and numerical description of the transition process of continuous internal nozzle flow to dispersed phase flow outside of the nozzle. However, the primary breakup mechanisms are still not sufficiently understood yet due to a lack of experimental data for the atomisation region. Classical breakup models like the Taylor Analogy Breakup (TAB) model [129], the model of Reitz and Diwakar [145], and the Wave model [146] do not distinguish between the two breakup mechanisms. The parameters of these models are tuned to match experimental data further downstream, in the region of secondary breakup. Other models, like the Enhanced Taylor Analogy Breakup (ETAB) model [172] or the Kelvin Helmholtz-Rayleigh Taylor (KH-RT) model [148] treat the primary breakup region separately to simulate both breakup processes independently. The correct values for the set of primary breakup parameters, however, are not easy to determine due to the lack of experimental data.

For this reason, in this work the liquid atomisation is not modelled based on a physical approach, it is rather prescribed by using a droplet Size Distribution Function (SDF). The droplet SDF is accurately validated against pressure chamber measurements, excluding uncertainties due to poor secondary breakup modelling, by setting boundary conditions so that measurements are correctly reproduced.

Following the approach of Abad Lozano [1], a $log-normal$ probability density function $F(d_d)$ is used to describe the droplet size distribution:

$$F(d_d) = \frac{1}{\sqrt{2\pi}\,\sigma_d} \cdot e^{\left[-\left((\ln(d_d)-\ln(\bar{d}_d))^2\right)/\left(2\cdot\sigma_d^2\right)\right]} \tag{2.38}$$

where $\bar{d}_d$ is the mean diameter and σ_d the droplet size distribution variance. The mean diameter is varied only in order to match the spray penetration, while leaving the standard deviation σ_d to a constant value of 0.5.

2.3.4 Droplet Breakup

The (secondary) breakup of the droplets into smaller, more stable fragments, is determined by aerodynamic forces resulting from the relative movement between the droplet and the gas phase. Thereby, the driving force increases proportionally with the relative velocity $\mathbf{v}_r$. Whereas, an increased surface tension and viscosity[1] of the liquid droplet counteract the breakup process. The ratio of both parameters determines the droplet deformation, and is expressed as non-dimensional Weber number

---

[1] An increased viscosity retards the deformation and initiated oscillation of the droplet.

We_d.

$$\text{We}_d = \frac{\rho \, |\mathbf{v}_r|^2 \, d_d}{2\gamma_d} \tag{2.39}$$

Dependent on the Weber number, Nicholls [127] identified two different breakup regimes, the bag breakup and the stripping breakup. Bag breakup occurs when a high pressure exerted on the front of a droplet leads to a bag shaped deformation of a droplet and eventually breakup into smaller droplets. Stripping breakup happens when the edge of a bag shaped droplet is stripped into smaller droplets due to a shear by a relative velocity between the droplet and the ambient. Patterson and Reitz [130] suggested that the Rayleigh-Taylor normal surface wave is responsible for the bag breakup, while the Kelvin-Helmholtz surface wave is for the stripping breakup[1]. However, these two mechanisms compete simultaneously.

With regard to the breakup regimes defined by Nicholls, Reitz and Diwakar [147] developed a breakup model, which describes the rate of change of droplet diameter during the continuous droplet breakup as rate expression

$$\frac{\mathrm{d}d_d}{\mathrm{d}t} = -\frac{d_d - d_{d,stable}}{\tau_{d,b}} \tag{2.40}$$

where d_d is the instant droplet diameter, $\tau_{d,b}$ the characteristic droplet lifetime and $d_{d,stable}$ the maximal stable droplet diameter defined as

$$d_{d,stable} = -\frac{\gamma_d}{\rho_d^2 \, v_r^3 \, \nu_d}. \tag{2.41}$$

The corresponding breakup time $\tau_{d,b}$, i.e. the time between droplet deformation and fragment formation, needs to be modelled.

$$\tau_{d,b} = \begin{cases} C_b \sqrt{\left(\rho_d \left(d_d/2\right)^3\right)/\gamma_d} & \text{if } \text{We}_d > 6 \text{ (Bag Breakup)} \\[2ex] C_s \left(d_d/2\right)/v_r \sqrt{\rho_d/\rho_g} & \text{if } \text{We}_d/\sqrt{\text{Re}_d} > 0.5 \text{ (Stripping Breakup)} \end{cases} \tag{2.42}$$

The constants are defined to $C_b = \pi$ and C_s is of order unity.

---

[1]Rayleigh-Taylor normal acceleration surface instability develops and breaks a large droplet into small droplets. A Kelvin-Helmholtz surface tension wave grows and rips small droplets from the parent droplet.

2.3.5 Droplet Turbulence Interaction

A droplet traveling through a turbulent flow field will be deflected from
its course due to the action of randomly varying velocity fluctuations.
Using a stochastic approach [161], it is assumed that the droplet velocity is for a characteristic droplet turbulence interaction time randomly
affected by the turbulence. The characteristic droplet turbulence interaction time depends on the minimum of the eddy lifetime and the transit
time required for the droplet to cross the eddy. Latter one is calculated
from the linearised equation of motion for a particle in a uniform flow.

$$t_t = -\delta t \cdot \ln \left(1 - \frac{l_t}{\delta t \cdot |\mathbf{v}_r|} \right) \tag{2.43}$$

When $l_t > \mathbf{v}_r \, \delta t$, the linearised stopping distance of the particle is
smaller than the characteristic length scale of the eddy and the equation
has no solution. In this case, the eddy has captured the particle and the
interaction time is the eddy lifetime.

2.3.6 Spray Wall Impingement

If a droplet impinges a wall, the droplet can either stick to the wall forming wall film, rebound, spread, or splash. The result of the impingement
depends on the characteristics of the droplet, the impingement surface
and the fluid transporting the droplet to the wall. For classification of
impingement mode, the Weber number of the impinging droplet $We_{d,in}$
is used [10, 165]. The corresponding allocation is displayed in table 2.1.

The critical impingement Weber numbers $We_{d,cr,wet}$ and $We_{d,cr,dry}$
are determined by the droplet Laplace number $La = \rho_d \, \gamma_d \, d_d / \mu_d^2$, and in
addition to for dry walls [168] by the surface roughness Δ_{surf}.

$$We_{d,cr,wet} = 1320 \cdot La^{-0.18} \tag{2.44}$$

$$We_{d,cr,dry} = \begin{cases} 5264 \cdot La^{-0.18} & \text{if } \Delta_{surf} \leq 0.05 \, \mu m \\[2mm] 2629.536 / \Delta_{surf}^{0.24} \cdot La^{-0.18} & \text{if } 0.05 \, \mu m < \Delta_{surf} \leq 12 \, \mu m \\[2mm] 1322 \cdot La^{-0.18} & \text{if } \Delta_{surf} > 12 \, \mu m \end{cases}$$

$$\tag{2.45}$$

Wet Wall	
Stick	$We_{d,in} < 5$
Rebound	$5 \leq We_{d,in} < 10$
Spread	$10 \leq We_{d,in} < We_{d,cr,wet}$
Splash	$We_{d,in} \geq We_{d,cr,wet}$
Dry Wall	
Stick	$We_{d,in} < We_{d,cr,dry}$
Splash	$We_{d,in} \geq We_{d,cr,dry}$

Table 2.1: Classification of spray wall impingement mode for wet and dry walls according to [10, 165]

If the wall temperature is greater than the Leidenfrost temperature T_L, a thin vapour cloud prevents droplets from contacting the wall. The Leidenfrost temperature is defined according to the impingement direction to the surface:

$$
T_L = \begin{cases}
T_{d,boi} + 135.6 \cdot We_{d,in}^{0.09} & \text{if normal} \\
& \text{impingement} \\[2ex]
T_{d,boi} + \left(\left(T_{d,boi} + 135.6 \cdot We_{d,in}^{0.09} \right) - T_{d,boi} \right) & \text{if oblique} \\
\cdot \left(0.28 \cdot 180\alpha/\pi - 0.0019 \cdot (180\alpha/\pi)^2 \right) & \text{impingement}
\end{cases}
$$

$$(2.46)$$

where α is the incidence angle of the droplet relative to surface normal direction [193], and $T_{d,boi}$ is the droplet boiling temperature. In analogy to the normal impingement, the impingement mode in the Leidenfrost regime is determined by the Weber number of the impinging droplet [159], as illustrated in table 2.2.

Rebound	$We_{d,in} < 50$
Secondary breakup in vertical direction	$50 \leq We_{d,in} < 80$
Secondary breakup in horizontal direction	$We_{d,in} \geq 80$

Table 2.2: Classification of spray wall impingement mode in Leidenfrost regime according to [159]

A droplet that rebounds from the wall retains mass and diameter. The outgoing tangential and normal velocities are evaluated according to [111]

$$v_{d,out}^t = \frac{5}{7} \cdot v_{d,in}^t \qquad (2.47)$$

$$v_{d,out}^n = 0.993 - 1.76 \cdot \alpha + 1.56 \cdot \alpha^2 - 0.49 \cdot \alpha^3 \, v_{d,in}^n. \qquad (2.48)$$

In case droplet splash occurs, a part of the droplet mass deposits to the surface film and the remaining part of the droplet mass is returned as secondary breakup droplet. The mass of the secondary breakup droplet $m_{d,sec}$ is defined as [10]

$$m_{d,sec} = \begin{cases} m_{d,in} \left(0.2 + 0.6 \cdot N_{ran}\right) & \text{if dry wall} \\[2ex] m_{d,in} \left(0.2 + 0.9 \cdot N_{ran}\right) & \text{if wet wall} \end{cases} \qquad (2.49)$$

where $m_{d,in}$ is the mass of the impinging droplet and N_{ran} a random number ($0 \leq N_{ran} \leq 1$). The number of secondary droplets is defined as [10]

$$N_{d,sec} = 5 \cdot \left(\frac{\text{We}_{d,in}}{\text{We}_{d,cr}} - 1\right). \qquad (2.50)$$

The diameter of the secondary breakup droplets can be determined from the mass conservation[1].

$$d_{d,sec} = \sqrt[3]{\frac{6 \cdot m_{d,sec}}{N_{d,sec} \, \pi \, \rho_{d,sec}}} \qquad (2.51)$$

The velocity of the secondary breakup droplet can be evaluated from the energy conservation, with

$$v_{d,sec} = \sqrt{\frac{2 \cdot \left(E_{d,sec,tot} - E_{d,sec,surf}\right)}{m_{d,sec}}} \qquad (2.52)$$

where $E_{d,sec,tot} = E_{d,in,tot} - E_{d,sec,splash}$ is the total energy of the secondary breakup droplet with $E_{d,sec,splash} = \text{We}_{d,cr}/12 \, \pi \, d_d^2$ as energy required in the splashing process, and $E_{d,sec,surf} = \pi \, \gamma_d \left(N_{d,sec} \, d_{d,sec}^2\right)$ as sum of the secondary breakup droplets surface energy.

The size of the secondary breakup droplet in the Leidenfrost regime is dependent on the Weber number of the impinging droplet [159]. For

---

[1]Assuming the injected mass is determined.

the vertical secondary breakup the droplet size is defined as

$$d_{d,sec} = d_{d,in} \left(1.07 + 0.0101 \cdot \mathrm{We}_{d,in} + 0.0000329 \cdot \mathrm{We}_{d,in}^2 \right) \qquad (2.53)$$

and for the horizontal secondary breakup as

$$d_{d,sec} = \begin{cases} d_{d,in}(1.07 + 0.0101 \cdot \mathrm{We}_{d,in} & \text{if } \mathrm{We}_{d,in} \leq 140 \\ \quad +0.0000329 \cdot \mathrm{We}_{d,in}^2) & \\ \\ d_{d,in} \cdot 0.416 \cdot 10^{-0.00102 \cdot \mathrm{We}_{d,in}} & \text{if } 140 < \mathrm{We}_{d,in} \leq 300 \\ \\ d_{d,in} \cdot 0.2 & \text{if } \mathrm{We}_{d,in} > 300. \end{cases}$$
$$(2.54)$$

The normal velocity of the outgoing droplet equals 1/9-th of the normal velocity of the impinging droplet. The tangential velocity can be determined from momentum conservation.

2.3.7   Wall Film Modelling

The liquid fuel remaining on the surface formed by impinging spray needs to be modelled as wall film. For modelling of wall film, Bai and Gosman [11] proposed a model, which adopts the Eulerian approach and seeks for the solution of the wall film characteristics by solving continuous wall film transport equations. Assuming a sufficiently thin wall film, the wall film transport equations can be cast into depth-averaged equations, which are two-dimensional in sense of that only a surface mesh needs to be considered. To discretise the wall film transport equations, a finite volume method is applied. The method involves the subdivision of the wall film flow domain in a finite number of smaller, non-overlapping control volumes, so-called patch cells P. Thereby, a depth-averaged quantity is defined as

$$\overline{\Phi} = \frac{1}{l_{Wf}} \int_0^{l_{Wf}} \Phi \, \mathrm{d}x_Z \qquad (2.55)$$

where l_{Wf} is the wall film thickness and x_Z the distance from the surface normal. The underlying conservation equations will be outlined in the following briefly. For simplicity, the bar is omitted for all depth averaged film properties in the expressions of the discretised film.

Continuity

The discretised continuity equation reads

$$\frac{\left((\rho_{wf}\, A\, l_{Wf})_P^{t+\Delta t} - (\rho_{wf}\, A\, l_{Wf})_P^{t}\right)}{\Delta t} + \sum_{i=1}^{N_{P,f}} l_{wf,P,f}\, l_{P,f}\, F_{P,f} \qquad (2.56)$$

$$= \dot{m}_{dep} + \dot{m}_{vap}.$$

The summation $\sum_{i=1}^{N_{P,f}}$ is done over all sides f of the patch cell P. The term $l_{P,f}$ denotes the length of the patch cell side f, $F_{P,f} = \rho_{wf}\, \mathbf{v}\, \mathbf{n}_{lf}$ the film face flux at side f with $\mathbf{n}_{lf}$ as unit vector tangentially normal to side f, and $\dot{m}_{dep}$ and $\dot{m}_{vap}$ the net droplet mass deposition rate and the net film evaporation rate at interface, respectively.

Momentum

The momentum equation is given by

$$\frac{\left((\rho_{wf}\, \mathbf{v}\, A\, l_{Wf})_P^{t+\Delta t} - (\rho_{wf}\, \mathbf{v}\, A\, l_{Wf})_P^{t}\right)}{\Delta t} \qquad (2.57)$$

$$+ \sum_{i=1}^{N_{P,f}} l_{wf,P,f}\, l_{P,f}\, F_{P,f}\, \mathbf{v}_{P,f}$$

$$= - \sum_{i=1}^{N_{P,f}} p_{P,f}\, l_{wf,P,f}\, l_{P,f}\, \mathbf{n}_{lf} + \mu_{wf} \left.\frac{\partial \mathbf{v}_t}{\partial x_Z}\right|_{int} - \mu_{wf} \left.\frac{\partial \mathbf{v}_t}{\partial x_Z}\right|_{w} + \omega_{Spray,\mathbf{v}}$$

where $p_{P,f}$ is the total film pressure equaling the sum of impingement pressure, gas pressure and capillary pressure, the term $\left.\frac{\partial \mathbf{v}_t}{\partial x_Z}\right|_{int}$ is the shear stress at the film interface, $\left.\frac{\partial \mathbf{v}_t}{\partial x_Z}\right|_{w}$ is the shear stress at the wall with $\mathbf{v}_t$ as film tangential resultant velocity, and $\omega_{Spray,\mathbf{v}}$ the droplet tangential momentum source.

Energy

The enthalpy equation reads

$$\frac{\left((\rho_{wf}\, h\, A\, l_{Wf})_P^{t+\Delta t} - (\rho_{wf}\, h\, A\, l_{Wf})_P^{t} \right)}{\Delta t} + \sum_{i=1}^{N_{P,f}} l_{wf,P,f}\, l_{P,f}\, F_{P,f}\, h \quad (2.58)$$
$$= \dot{q}_{int} - \dot{q}_w + \omega_{Spray,h}$$

where $\dot{q}_{int}$ is the heat flux at film interface and $\dot{q}_w$ the heat flux at wall. The term $\omega_{Spray,h}$ represents the droplet enthalpy source.

Chapter 3

Flame Propagation Modelling

3.1 Introduction

Combustion processes can be classified by their flame type into premixed
or diffusion flames [134]. Diffusion flames occur under non-premixed
conditions, as they appear in conventional diesel engines, where fuel and
air are mixed by convection and diffusion during combustion. Optimal
combustion conditions are on the surface of stoichiometric mixture, i.e.
a mixture ratio where fuel and air would be both consumed entirely. As
the combustion process is faster than the diffusion process, diffusion is
the rate limiting step.

Premixed flames occur in conventional SI engines as the combustion
starts out under premixed conditions. The premixed flame, where the
chemical reaction occurs, is relatively thin (typically in the order of
Zeldovich thickness $0.1 \leq l_f \leq 1$ mm [176]) and separates the unburnt
and the burnt gases region [171], as illustrated in figure 3.1.

In absence of turbulence, the premixed flame would propagate towards fresh gases with a laminar flame speed s_l, which depends on fuel
type.

However, in SI engines a turbulent flow field is inducted by high
shear flows in the intake manifold [70]. The turbulence has a great
impact on the premixed flame and the flame propagation is therefore
determined by the turbulent flame speed s_t rather than by the laminar
flame speed. Turbulent eddies can wrinkle and strain the flame surface, enhancing the flame surface area, whose thickness can rather be
described by the turbulent flame thickness than by the laminar flame
thickness. But the inner flame layer, where the fuel is consumed, is
assumed to not perturbed by turbulent eddies [134].

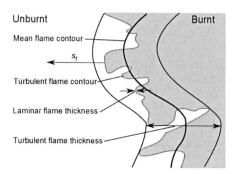

Figure 3.1: Schematical illustration of premixed turbulent flame front according to [191]

In order to model the premixed turbulent combustion process, different modelling approaches are proposed in the literature. A brief survey of these models is given in section 3.2. For modelling of premixed flame propagation bases in this work on the G-equation approach, which is introduced in section 3.3. The implementation in the 3D-CFD code is described in section 3.4. The model validity for turbocharged direct injection SI engines is proven in section 3.5. The G-equation gives a kinematic description of the flame front involving the displacement speed s_t, which is determined by the turbulence and the laminar flame speed s_l, and needs to be modelled. Different closure formulations for the laminar flame speed and the turbulent flame speed are investigated in sections 3.6 and 3.7. Afterwards, the combustion model is validated using optical measurements in section 3.8.

3.2   Literature Survey

Most combustion models proposed in the literature are derived from non-premixed combustion. Under non-premixed conditions, the flame front in the combustion chamber is represented by the stoichiometric mixture fraction[1]. This enables the prediction of combustion, emission formation, as well as flame structure by an auto ignition based chemistry model coupled with 3D-CFD work flow [162]. However, this modelling approach fails to predict combustion under premixed conditions, where

---

[1]The mixture fraction Z represents in a separated flow of fuel and oxidizer the mass fraction of fuel in the mixture [134].

the burning rate is rather controlled by turbulence than by molecular diffusion of reactants toward the reaction zone.

For the description of the turbulent premixed flame a progress variable c is used, which is defined for a simple one-step irreversible chemical reaction, i.e. reactant $\rightarrow$ product, as [176]

$$c = \frac{T - T_u}{T_b - T_u} \qquad \text{or} \qquad c = \frac{Y_F - Y_F^u}{Y_F^b - Y_F^u} \tag{3.1}$$

where T denotes the local temperature, T_u the unburnt temperature, and T_b the burnt temperature, such as $c = 0$ refers to the fresh gases and $c = 1$ to the fully burnt gases. The definition of the progress variable based on the fuel mass fraction Y_F is in analogous manner. The flame front position is assumed to correspond to values of the progress variable.

The most widely used turbulent premixed combustion models represented in the literature either base on geometrical analysis or the turbulent mixing approach [176]. The most important ones are listed in table 3.1 along with the modelling tools used and issues involved. Another type of analysis bases on statistical description of the turbulent premixed combustion. The so-called PDF approaches are excluded from discussion below.

Combustion models which base on geometrical analysis are usually linked to the flamelet assumption[1], enabling the decoupling of chemistry and turbulent flow modelling. The description of the flame front structure can either base on the study of a scalar field in terms of dynamics and physical properties of iso-value surfaces of the progress variable defined as flame surfaces, or by gathering information in the direction normal to the flame surface by focusing on the structure of reacting flow [176]. Corresponding formulations proposed are the G-field, flame surface density concepts, and the flame surface wrinkling approach.

The G-equation [134] provides a kinematic description of the flame front, where the mean turbulent flame brush is located at $\widetilde{G} = G_0$. The kinematic description involves the displacement speed s_t of the turbulent flame front, which needs to be modelled. By considering the turbulent flame as a propagating surface, the internal flame structure does not need to be resolved. Moreover, the G-equation simulates comparatively

---

[1]Assuming that the inner flame layer can not be perturbed by the turbulence and following the chemical time scales are smaller than the convective and diffusive time scales, the flame can be considered as an ensemble of thin reactive-diffusive layers, called flamelets, which are embedded in the turbulent flow [134].

thin flames [93], enabling the modelling of auto-ignition phenomena in the unburnt mixture.

Description	Tools	Modelling issues	Models	Reference
Geometrical	**G-field** with $G = G_0$ at the flame, $\partial \widetilde{G}/\partial t + \widetilde{\mathbf{v}}\nabla\widetilde{G}$ $= \rho_u/\overline{\rho}\, s_t |\nabla\widetilde{G}|$	s_t	G-Equation	[134]
	Flame surface density Σ Algebraic closure or transport equation	$\overline{\omega}_i = \dot{\Omega}_i\Sigma,$ Total stretch =curvature +strain rate	BML CPB CFM ECFM ECFM-3Z MB CD CH	[25] [27] [44] [45] [36] [108] [32] [34]
	Flame surface wrinkling Ξ	$\overline{\Xi} = \Sigma/|\nabla\overline{c}|$	Weller	[181]
Turbulent mixing	**Scalar dissipation rate** χ $\overline{\rho\widetilde{\chi}} = \overline{\rho D |\nabla Y|^2}$	$\overline{\omega} \approx \widetilde{\chi},$ Rate of micromixing	EBU EDM	[164] [107]

Table 3.1: Tools for turbulent premixed combustion modelling (according to [176]) and corresponding combustion models

In flame surface density concepts the flame is defined as a surface by introducing the flame surface density Σ [176]. The mean burning rate $\overline{\omega}_i$ of a species i is modelled as product of mean local burning rate $\dot{\Omega}_i$ and the flame surface density Σ. Former parameter is related to the properties of the local flame front and can be estimated from a prototype laminar flame. For determining Σ, algebraic relations or balance equations are used. For the solution of the flame front area Bray et al. [25] proposed an algebraic expression, which is due to its similarity often referred as Bray-Moss-Libby (BML) model. A balance equation for the solution of the flame front area was first proposed by Marble and Broadwell [109]. Their model bases on phenomenological analysis and was derived for modelling of non-premixed turbulent combustion. For the use in premixed turbulent combustion, a multitude of flame surface density balance equation closures have been derived over the past decades. These closure formulations are by name the model of Cant, Pope and Bray (CPB) [27], the Coherent Flame Model (CFM) [44] and its derivatives Extended Coherent Flame Model (ECFM) [45] and 3-Zone Extended Coherent Flame Model (ECFM-3Z) [36], the model of Mantel and Borghi (MB) [108], the model of Cheng and Diringer (CD) [32], and the model of Choi and Huh (CH) [34]. The formulations differ in terms of modelling of the strain rate acting on surface induced by the mean flow flied, the stain rate due to turbulence motion, and the consumption of the flame area.

The flame surface density formalism can be recast in term of flame surface wrinkling [176] by introducing the ratio of the flame surface to its projection, the so-called wrinkling factor Ξ. A transport equation for Ξ can be derived from those of the progress variable and the surface density [182]. However, compared with surface density formulations, the resulting equations are more complicated [176].

In general, flame surface density and flame surface wrinkling models define the flame propagation in an implicit way. As a consequence, for prediction of mean heat release and emission formation these models demand for a detailed resolution of the flame front [93]. Additionally, these models simulate comparatively thick turbulent flames [93], disabling auto-ignition modelling in the unburnt mixture.

Another form of flame analysis is the turbulent mixing approach [176]. The underlying assumption is the control of reactant mixing and thus the reaction rate by small-scale dissipation of species. The molecular mixing is thereby quantified using the scalar dissipation χ of the progress variable. For the description of the reaction rate, Spalding [164] derived the Eddy Break-Up (EBU) model, which bases on phenomenological analysis of the turbulent combustion. The model describes the mean reaction rate as mainly controlled by turbulent mixing time τ_t (which is related to ε), and fluctuations of the progress variable, which characterise non-homogeneity and intermittencies. Latter one needs to be modelled. Since the EBU model is based on a global reaction equation, its application is limited to stoichiometric conditions. To mimic chemical features, the modelling constant needs to be tuned. Magnussen [107] extended the EBU model for the use under fuel rich and fuel lean conditions. The so-called Eddy Dissipation Model (EDM) bases on a global reaction equation, too.

The oftentimes highlighted deficits of the EBU modelling approaches are the over-estimation of the reaction rate, especially in highly strained regions, where ε/k is large (flame-holder weaks, walls, etc.) [97, 176], as well as the over-prediction of temperature and species concentrations [77]. Additionally, the mathematical characteristics of the EBU models are comparable to the one of the flame area models, requiring a highly resolved flame front area [77]. However, the modelling approach distinguishes with easy implementation, simplicity, and steady converge.

Summarising, flame area evolution models, i.e. flame surface density and flame surface wrinkling models, as well turbulent mixing models define the flame propagation in an implicit way. Whereas, the flame propagation is described in the G-equation approach in an explicit way, reducing the dependency on highly resolved flame front area [93]. Moreover, in contrast to other modelling approaches, the G-equation simulates comparatively thin flames [93], enabling the modelling of auto-ignition phenomena in the unburnt mixture. These fact makes the G-equation to an appealing solution for modelling of premixed turbulent combustion.

3.3 G-Equation Model Description

The G-equation model bases on the flamelet assumption, which assumes that the inner flame layer can not be perturbed by the turbulence and following the chemical time scales are smaller than the convective and diffusive time scales. In this way the flame can be considered as an ensemble of thin reactive-diffusive layers, called flamelets, which are embedded in the turbulent flow [134]. With its application to combustion modelling, Williams [190] first suggested the use of a transport equation for a non-reactive scalar G for laminar flame propagation. Peters [132] subsequently extended this approach to the turbulent flame regime.

In order to track the flame front, Peters [134] derived a field equation of the scalar $G(\mathbf{x}, t)$, where the mean flame front is located at [134]

$$G(\mathbf{x}, t) = G_0. \tag{3.2}$$

This interface divides the flow field into an unburnt region $(G < G_0)$ and a burnt gas region $(G > G_0)$. The differential equation describing the mean flame front position is derived by differentiation of equation 3.2 with respect to t.

$$\frac{\partial G}{\partial t} + \nabla G \cdot \frac{\mathrm{d}\mathbf{x_f}}{\mathrm{d}t} = 0 \tag{3.3}$$

The change of the mean flame front position $\mathrm{d}\mathbf{x}_f/\mathrm{d}t$ is assumed to depend on the mean flow velocity $\overline{\mathbf{v}}$ and the turbulent flame propagation speed s_t.

$$\frac{\mathrm{d}\mathbf{x}_f}{\mathrm{d}t} = \overline{\mathbf{v}} + \mathbf{n}\, s_t \tag{3.4}$$

Herein, $\mathbf{n}$ denotes the vector normal to the flame front pointing towards the unburnt mixture.

$$\mathbf{n} = -\frac{\nabla G}{|\nabla G|} \tag{3.5}$$

Introducing 3.4 in 3.3, the G-equation reads

$$\frac{\partial G}{\partial t} + \overline{\mathbf{v}} \cdot \nabla G = s_t\, |\nabla G|. \tag{3.6}$$

The implementation of the G-equation in the 3D-CFD code and the coupling with the progress variable are described in the next section.

3.4   Implementation in 3D-CFD Code

For numerical solution, the G-equation is written as the standard finite-volume transport equation [29]

$$\int_V \frac{\partial \rho G}{\partial t}\,\mathrm{d}V + \int_\Gamma \rho\, v\, G\,\mathrm{d}\Gamma = \int_V \rho\, s_t\,\mathrm{d}V \tag{3.7}$$

where V denotes the cell volume and Γ the cell bounding surface. Until the start of combustion the G-equation is inactive and the initial G value is set to -10. As soon as the combustion starts, the source term on the right hand side of equation 3.7 needs to be calculated.

In order to evaluate the source term in the initial flame kernel development phase, it is assumed that the premixed flame propagates from the spark plug as a spherical surface. With increasing kernel size, the initial flame kernel propagation speed s_{kern} grows from the laminar flame speed to the turbulent flame speed.

$$s_{kern} = s_l + \alpha_{kern}\left(s_t - s_l\right) \tag{3.8}$$

Herein, α_{kern} describes the ratio of current flame kernel diameter d_{kern}, which is determined for each time step according to

$$d_{kern,new} = d_{kern,old} + s_{kern}\Delta t, \tag{3.9}$$

and a critical kernel diameter $d_{kern,Burnup}$, which is an engine specific calibration parameter [21].

$$\alpha_{kern} = \left(\frac{d_{kern}}{d_{kern,Burnup}}\right)^{\beta_{kern}} \tag{3.10}$$

The flame kernel development phase is completed as soon as the criteria $d_{kern} \geq d_{kern,Burnup}$ is fulfilled, and the transport equation 3.7 is solved.

In order to determine whether the flame front reaches a cell or not, i.e. to define the turbulent flame thickness or flame brush thickness, Peters [134] derived a transport equation for the variance of G. Together with the mean value of G and with the Reynolds-Averaged Navier-Stokes equations and the turbulence modelling equations, the model forms a complete set to describe the turbulent flame front propagation [103]. However, in some SI engines the turbulent flame brush thickness is of the same order of the cell size in a typical 3D-CFD simulation mesh, i.e. in the range of 1 to 4 mm [171]. To capture the detailed turbulent flame brush a small cell size is required, which, in turn, would exceed an acceptable time frame of engine combustion simulation. For this reason, the detailed turbulent flame structure is ignored and the turbulent flame thickness is estimated from the largest value between the cell size V_{Cell} and the mixing length scale, evaluated using the $k - \varepsilon$ model.

$$l_{f,t} = max\left(V_{Cell}^{1/3}, C_f\, C_\mu \frac{k^{1.5}}{\varepsilon}\right) \tag{3.11}$$

Herein, C_f is the flame thickness scaling factor and C_μ a constant of the $k - \varepsilon$ model. Accounting for the turbulent flame thickness, areas with $G \leq 0.5 l_{f,t}$ describe the unburnt mixture and areas with $G \geq -0.5 l_{f,t}$ the burnt mixture [21]. In the range $-0.5 l_{f,t} < G < 0.5 l_{f,t}$ the so-called reaction zone is found.

The G-equation provides a kinematic description of the flame front involving the displacement speed s_t. However, the reactant consumption and heat release rate are controlled by the consumption speed, which is described by the progress variable c. A widely used approach to model the progress variable c bases on the assumption, that with the propagation of the flame front, the mixture within the mean flame brush tends to local and instantaneous thermodynamic equilibrium [103,171]. However, this procedure can cause numerical instabilities due to the sharp gradient between the unburnt and burnt zone. In order to smooth the gradient, a burning rate behind the flame front is defined. As soon as the flame front reaches a cell, i.e. the G value is greater than the negative value of the half flame thickness $G > -0.5\, l_{f,t}$, the burning rate $\dot{m}_b$ is calculated by

$$\dot{m}_b = C_b \rho\, V_{Cell}^{2/3}\, s_t \qquad (3.12)$$

where C_b is a model constant, ρ the gas mixture density, and s_t the turbulent flame propagation speed[1]. Assuming the gas mixture is homogeneous in each cell, the change rate of the unburnt air, unburnt fuel, burnt air and burnt fuel mass are determined. The progress variable c is approximated by the relative ratio of burnt air and burnt fuel mass to total mass in a cell. Based on c, the reactant consumption and heat release rate are determined using the IPV model (see section 4.4).

3.5 Flamelet Model Validation for Turbo-Charged Direct Injection SI Engines

The G-equation model bases on the flamelet assumption, which implies that the inner flame layer can not be perturbed by the turbulence. The validity of this assumption for turbocharged direct injection SI engines is proven in the following by investigating the interaction between turbulence and chemistry.

In order to describe the interaction between turbulence and chemistry, the associated characteristic scales are introduced and related to non-dimensional numbers. The non-dimensional numbers are taken as basis to characterise the regimes of turbulent premixed combustion. The relevant turbulent premixed combustion regimes are identified for turbocharged direct injection SI engines. Afterwards, the interaction be-

---

[1]Note that for $d_{kern} < d_{kern,Burnup}$, in equation 3.12 the term s_t is replaced by s_{kern}.

tween turbulence and chemistry in a fully combustion process is investigated in detail.

3.5.1 Characteristic Scales of Turbulence and Chemistry

In order to characterise the interaction between turbulence and chemistry, the corresponding characteristic scales are introduced in the following and related in non-dimensional numbers.

Turbulence Scales

The characteristic of a system is determined by large-scale and small-scale motions. Large-scale motions are described by the turbulent velocity u' (given by equation 2.21), and the integral length scale l_t (given by equation 2.20). The integral length scale characterises the energy-containing eddies which are typically in the magnitude of geometrical dimension [56]. The smallest turbulent motions are characterised by the Kolmogorov velocity

$$v_k = (\nu \bar{\varepsilon})^{1/4} \qquad (3.13)$$

and Kolmogorov length l_k (given by equation 2.2) and are a characteristic of the dimension at which dissipation occurs [56].

Chemistry Scales

A premixed laminar flame can be characterised by the laminar flame velocity s_l, Zeldovich thickness l_f, and inner layer thickness l_δ [135].

$$l_f = \frac{D_f}{s_l} = \frac{\lambda/c_p}{\rho\, s_l} \approx \frac{0.0000258 \left(\frac{T}{298K}\right)^{0.7}}{\rho\, s_l} \qquad (3.14)$$

$$l_\delta = \delta\, l_f \qquad \text{with} \qquad \delta \approx 0.1 \qquad (3.15)$$

The different scales of a laminar flame are illustrated in figure 3.2.

The inner layer thickness is smaller than the thickness of the adjacent Zeldovich layer, which is in turn small compared to the chemically inert preheat zone upstream of the inner layer, indicating chemistry takes place around and above T_δ only [60].

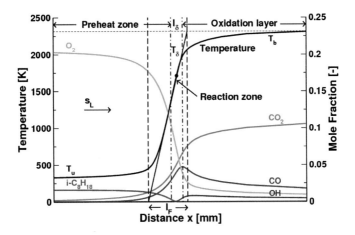

Figure 3.2: Calculated structure of a stoichiometric *iso*-Octane / air flame, layers classified according to [132]

Characteristic Scales related to Non-Dimensional Numbers

In order to describe the interaction between the turbulence and chemistry scales, non-dimensional numbers are introduced in the following.

The ratio of large-scale motions and chemical scales is used to determine whether combustion takes place in turbulent or laminar mode, expressed by the turbulent Reynolds number Re_t [135]. Assuming $Sc = \nu/D = 1$, the kinematic viscosity ν equalises the diffusivity D, and the turbulent Reynolds number reads

$$Re_t = \frac{u' l_t}{s_l l_f}. \tag{3.16}$$

The ratio of large-scale motion time scale $\tau_t = l_t/u'$ and chemistry time scale $\tau_c = l_f/s_l$ is used to determine whether chemistry is considered as fast or slow [135], and is expressed as turbulent Damköhler number

$$Da = \frac{\tau_t}{\tau_c} = \frac{s_l \, l_t}{u' \, l_f}. \tag{3.17}$$

For high Damköhler numbers, the chemistry is considered fast, i.e. the reaction time is short. In this case, reaction sheets of various wrinkled types can occur. For low Damköhler numbers, the chemistry is considered slow and the system can be described as a well-stirred flow [56].

To characterise the influence of flame stretch due to turbulent flow field, the time scale of the smallest eddies $\tau_k = (\nu/\bar{\varepsilon})^{0.5}$ is related to the chemical time scale τ_c in the turbulent Karlovitz number [135]

$$\text{Ka} = \frac{\tau_c}{\tau_k} = \frac{l_f^2}{l_k^2} = \frac{v_k^2}{s_l^2}. \tag{3.18}$$

To describe the influence of the flow field on the inner reaction zone, the length scale of the smallest eddies of the turbulent flow l_k is related to the inner layer thickness l_δ in the Karlovitz number [135]

$$\text{Ka}_\delta = \frac{l_\delta^2}{l_k^2}. \tag{3.19}$$

3.5.2   Regime Diagram of Turbulent Premixed Combustion

In order to illustrate the interaction between turbulent flow field and laminar flame, Borghi introduced a diagram, which was later on further developed by Peters [134]. The so-called Borghi-Peters diagram is displayed in figure 3.3 and depicts the ratio of turbulent flow to laminar flame in terms of length scale and velocity, i.e. the ratios l_t/l_f and u'/s_l.

Based on the non-dimensional numbers introduced in section 3.5.1, Peters [132] identified in his work five different regimes characterising the premixed turbulent combustion. Those are listed in the order of increasing impact of the flow on laminar flame: laminar flames, wrinkled flamelets, corrugated flamelets, thin reaction zone and broken reaction zone. The regimes are outlined in the following starting from small ratios of u'/s_l and successively increasing u' by keeping the ratio l_t/l_f constant.

In the range $0.1 \leq u'/s_l \leq 1.0$ the flame speed is higher than the turbulent velocity, i.e. $s_l > u'$. Interpreting u' as turnover velocity of a large eddy, the flame propagation velocity is faster than the eddy turnover velocity. Hence, velocity fluctuations influence the laminar flame only marginal and the flames appear to be *wrinkled* [132].

Above the $u'/s_l = 1$ and underneath the Ka = 1 characteristic lines the *corrugated flamelets* regime is present. According to equation 2.20, an increase of u' by keeping l_t constant leads to an increase of the turbulent velocity of dissipation. Thus, the length of the smallest turbulent eddies decreases (see equation 2.2). However, in the flamelets regimes (Ka < 1, high Da-number), the Karlovitz numbers are smaller than one, i.e. the flame is still thinner than the length of the smallest turbulent eddy (see equation 3.18). Thus, the velocity of the largest eddies dominates the burning velocity of the flame. The large scale eddies push the flame front around causing a substantial corrugation. The smallest turbulent eddies featuring a smaller turnover velocity than flame velocity will not penetrate into the flame front. The flame stays quasi steady-state.

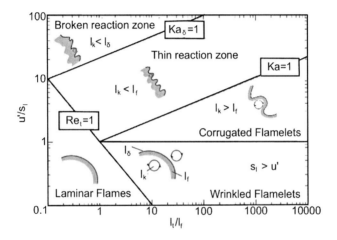

Figure 3.3: Regime diagram of turbulent premixed combustion according to [132]

Increasing the turbulent velocity further, the Ka = 1 characteristic line is passed and the *thin reaction zone* regime present, which is upper bounded by the $Ka_\delta = 1$ criteria. Ka greater than one indicates that the length of the smallest turbulent eddies is smaller than the flame thickness (see equation 3.18). As illustrated in figure 3.2, the preheating zone of the flame is much thicker than the inner reaction zone. The $Ka_\delta < 1$ criteria indicates, that the inner layer thickness is smaller than the size of the smallest eddies (see equation 3.19), holding:

$$u' > s_l < v_k \qquad \text{and} \qquad l_k > l_\delta. \tag{3.20}$$

Hence, the smallest turbulent eddies with the size l_k can enter into the preheat zone. The inner reaction zone will be crinkled by the turbulent eddies and locally large curvatures occur, i.e. the flame will be strained. But the inner reaction zone stays quasi steady-state.

Above the $Ka_\delta = 1$ characteristic line the broken reaction zone is present. The criteria indicates that the Kolmogorov length l_k is smaller than the thickness of the inner reaction layer l_δ, i.e. $l_k < l_\delta$. Hence, the smallest eddies can enter the inner reaction zone. Radicals which are required for the fuel decomposition may be passed outside of the inner reaction zone due to turbulent transport. As a result, a local extinguish of the inner reaction zone and following an extinguish of the whole flame front may occur.

Following this classification, the regimes characterising the premixed turbulent combustion where the flamelet assumption holds are the thin reaction zone regime, the corrugated flamelet regime and the wrinkled flamelets regime.

3.5.3 Regimes of Practical Interest for Turbocharged Direct Injection SI Engines

In the following, the regimes of turbulent premixed combustion introduced in section 3.5.2 are investigated in terms of practical interest in respect to turbocharged direct injection SI engines. The investigation bases on the different engines used for model validation in this work.

The highly turbocharged SI engine is used to validate the flame propagation model at part load Operating Point (OP). The same engine is used to validate the auto-ignition model in terms of pre-ignition phenomena prediction at full load OP. In order to prove the functionality of the auto-ignition model in terms of homogeneous auto-ignition processes, the HCCI engine operating at part load conditions is used. Additionally, in this study a mildly turbocharged engine is examined. For model validation, these engines are investigated in test bed measurements, all equipped with cylinder pressure indication, enabling thermodynamic analysis of the combustion process[1]. Additionally, optical measurement equipment is installed into the highly turbocharged SI engine.

For investigating the relevant regimes of turbulent premixed combustion, the upper extreme operating points of the engines are examined. The operating points are listed in table 3.2. The engine specific parameters can be found in appendix A.2.1, A.5.1, and A.3.1.

---

[1]The thermodynamic analysis provides the initial and boundary conditions for 3D-CFD analysis.

Engine	Operating point	Engine speed [rpm]	BMEP [bar]	ϕ [-]	ψ [-]
Highly turbo-charged engine	part load	2000	4.99	1.0	0.08
Mildly turbo-charged engine	part load	1500	5.0	1.0	0.09
HCCI engine	part load	1000	5.05	1.2	0.24

Table 3.2: Turbocharged direct injection SI engines and operating points used to examine relevant regimes of turbulent premixed combustion

Figure 3.4 displays the calculated characteristic engine curves in the regime diagram of turbulent premixed combustion of the different turbocharged SI engines, starting from point A (60 °CA bTDC) to point G (60 °CA aTDC).

The calculations base on test bed measurements of the different engines running fired at the operating points considered. However, for present investigation, the engines are examined under fired boundary conditions, but without spark ignition and thus combustion. The characteristic scales displayed are an average of the in-cylinder values determined using equations 2.20, 2.21, and 3.15, along with the laminar flame speed approximation function of Perlman introduced in section 3.6.

The characteristic curve of the highly turbocharged SI engine starts at point A (60 °CA bTDC) with a ratio of $u'/s_l = 8$ and $l_t/l_f = 37$. The ratio u'/s_l stays almost constant until point B (40 °CA bTDC), while the ratio of l_t/l_f increases continuously up to $l_t/l_f = 65$. In the following crank angle degrees until point D (TDC) the ratio l_t/l_f continues increasing up to $l_t/l_f = 100$, and the ratio u'/s_l starts to decrease reaching a value of $u'/s_l = 5$ at TDC. Until point E (20 °CA aTDC) the ratio u'/s_l continues decreasing reaching a value of $u'/s_l = 4$, which stays almost constant until point G (60 °CA aTDC). Starting from TDC the ratio l_t/l_f decreases, too, reaching a value of $l_t/l_f = 30$ at point G.

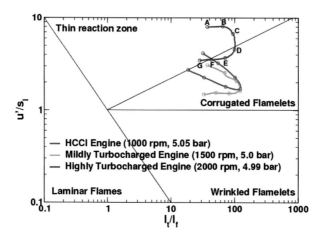

Figure 3.4: Calculated characteristic engine curves of non-fired tur-bocharged direct injection SI engines at operating points specified in table 3.2 in regime diagram of turbulent premixed combustion. Letters in the figure refer to A: 60 °CA bTDC, B: 40 °CA bTDC, C: 20 °CA bTDC, D: TDC, E: 20 °CA aTDC, F: 40 °CA aTDC, G: 60 °CA aTDC

Compared with the characteristic of the highly turbocharged SI engine, the characteristic curves of the HCCI engine and the mildly turbocharged SI engine possess in total lower values of the ratio u'/s_l and reach higher values of the ratio l_t/l_f at point D (TDC). Additionally, the rate of decrease of the ratio u'/s_l in the compression stroke (point A to D) is greater than the one of the highly turbocharged engine. In the expansion stroke (point D to G), the mildly turbocharged engine shows a comparable decrease of the ratio u'/s_l with the highly turbocharged engine. Whereas, the HCCI engine shows increasing values of u'/s_l.

In order to understand the impact of engine and operating point specific parameters on the characteristic engine curves in detail, the highly turbocharged SI engine is homogeneously initialised and different engine (integral length scale l_t) and operating point (turbulent velocity u', engine revolution speed RPM, temperature T, pressure p, fuel-air equivalence ratio ϕ, and EGR rate ψ) specific parameters varied separately. Figures 3.5 and 3.6 display the resulting characteristics in the regime diagram of turbulent premixed combustion.

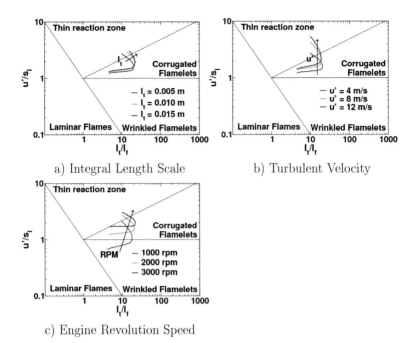

a) Integral Length Scale   b) Turbulent Velocity

c) Engine Revolution Speed

Figure 3.5: Calculated influence of engine and operating point specific turbulence scale parameters on characteristic engine curves in regime diagram of turbulent premixed combustion

Increasing the initial integral length scale (figure 3.5 a) from $l_t = 0.005$ m to $l_t = 0.015$ m the characteristic engine curves change primarily in the compression stroke in terms of increasing ratios of l_t/l_f as well as u'/s_l. An enhancement of l_t implies an increase of initial k (and thus u') and ε. Both variables decrease heavily in the compression stroke. Since the dissipation rate of ε is higher than the one of k, the integral length scale decreases, too. Thus, the effect of l_t on the characteristic engine curve is primarily limited to crank angle degrees before TDC. Hence, the interaction between the flame and turbulence will not be affected.

An increase of the initial turbulent velocity (figure 3.5 b) from $u' = 4$ m/s to $u' = 12$ m/s also results in a change of the characteristic engine curve primarily in the compression stroke in terms of increasing ratios of u'/s_l. Due to the continuous dissipation of k the ratio of u'/s_l is only marginally enhanced in the expansion stroke.

Another quantity strongly influencing the characteristic scales of the turbulence is the engine revolution speed (figure 3.5 c). An increase from 1000 to 3000 rpm results in an continuous enhancement of the ratio u'/s_l. The turbulent velocity is proportional to the piston speed which in turn is proportional to the engine revolution speed [70]. Due to an increasing u', the ratio of l_t/l_f is enhanced, too. Elevating the engine speed, the rate of dissipation of k and ε is decreased, resulting in a higher integral length scale.

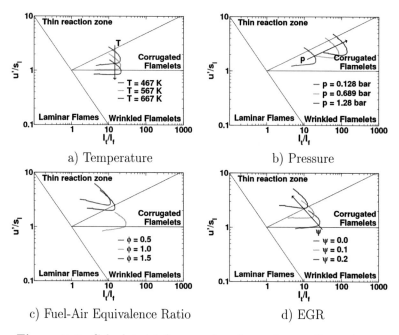

a) Temperature   b) Pressure

c) Fuel-Air Equivalence Ratio   d) EGR

Figure 3.6: Calculated influence of engine and operating point specific chemistry scale parameters on characteristic engine curves in regime diagram of turbulent premixed combustion

An enhancement of the initial in-cylinder temperature (figure 3.6 a) influences the ratio of u'/s_l particularly due to increasing laminar flame speed with increasing temperature. A minor effect on the flame thickness with respect to increasing ratio l_t/l_f can be observed. This is also a result of the increasing laminar flame speed.

Increasing the initial in-cylinder pressure (figure 3.6 b) from $p = 0.128$ bar to $p = 1.28$ bar results in an enhancement of the ratio u'/s_l as well as l_t/l_f. The first effect can be assigned to the decreasing flame speed with increasing pressure. Additionally to increasing pressure, the gas density enhances. Since the flame thickness is proportional to $1/\rho$ (see equation 3.15), l_f decreases with increasing p. As a result, the ratio of l_t/l_f is enhanced.

Enriching and enleaning the mean gas mixture (figure 3.6 c) up to $\phi = 1.5$ or alternatively $\phi = 0.5$, the ratio of u'/s_l increases while the ratio of l_t/l_f decreases. The enhancement of u'/s_l can be assigned to the decreasing laminar flame speed on the fuel rich and lean side, where $s_{l,\phi=0.5} < s_{l,\phi=1.5}$. The same is true for the decreasing ratio of l_t/l_f. Besides the proportionality of l_f to $1/\rho$, the flame thickness is also proportional to $1/s_l$. The decreasing density as a result of fuel enleaning has only a minor effect here.

The last parameter examined is the EGR rate (figure 3.6 d). Increased the mean rate from $\psi = 0.0$ to $\psi = 0.2$, the ratio of u'/s_l increases while the ratio of l_t/l_f decreases. With increasing EGR rate the laminar flame speed is reduced. As a result, the ratio of u'/s_l increases. Due to $l_f \propto 1/s_l$, the reduced laminar flame speed leads to a decrease of l_t/l_f here, too.

In the following, the results of the parameter variation are used to understand the engine specific characteristics of different engines and operating points displayed in figure 3.4. Starting at point A (60 °CA bTDC), the highest u'/s_l ratio possesses the highly turbocharged SI engine, the lowest ratio shows the mildly turbocharged engine. This can be ascribed to the different engine geometries, rotating speeds, as well as EGR rates of the engines and their operating points listed in table 3.2. The length scales of the engines defined by the geometry follow the ranking $l_{t,HCCI} > l_{t,mildly\,turbocharged} > l_{t,highly\,turbocharged}$. An increase of l_t results in an enhancement of the initial k (and thus u') and ε. The highly turbocharged engines operates at 2000 rpm, the mildly turbocharged engines at 1500 rpm, and the HCCI engine does at 1000 rpm. An increase of engine rotating speed results in an enhancement of u' and thus u'/s_l. Additionally, the HCCI engine operates with an internal EGR rate of $\psi = 0.24$, while the mildly and highly turbocharged engines operate with an EGR rate of $\psi = 0.09$ and $\psi = 0.08$, respectively. An increase of EGR leads to a decrease of s_l and thus to an enhancement of u'/s_l. In the following crank angle degrees until point D (TDC) the characteristic engine curves possess a different slope of u'/s_l. The slope can be ascribed to the engine specific length scales. Since the related k and ε and thus u' dissipate in the compression stroke and the dissipation rate of ε is higher than the one of k, the integral length scale decreases, too.

The characteristic engine curves differ additionally in the expansion stroke starting at point D (TDC) to point G (60 °CA aTDC). Since the mildly and highly turbocharged engines show a similar behaviour and only the HCCI engine differs heavily, the different behaviour can be ascribed to the high EGR rate of the HCCI engine. With increasing EGR rate the laminar flame speed is reduced and as a result the ratio of u'/s_l is enhanced.

Along general lines the characteristics of the mildly turbocharged SI engine and the HCCI engine are comparable to the one of the highly turbocharged SI engine. Considering the fact that combustion starts usually shortly before TDC, it can be stated that independently of engine examined the turbulent premixed combustion takes place in the regime of thin reaction zone and corrugated flamelets. Decreasing engine speeds or increasing temperatures can shift the combustion into the wrinkled flamelets regime. Thus, for turbocharged direct injection SI engines the flamelet assumption is fulfilled, enabling the modelling of premixed flame propagation with the G-equation approach.

3.5.4   Combustion Analysis in Regime Diagram of Turbulent Premixed Combustion

In section 3.5.3 the regime diagram of turbulent premixed combustion was investigated in terms of engine specific characteristics considering non-fired operation. In this section, the combustion process is analysed using the Borghi-Peters diagram. Figure 3.7 displays the characteristic curves of the fired and non-fired highly turbocharged engine at part load operating point (2000 rpm, 4.99 bar). As in section 3.5.3, the characteristic scales displayed are an average of the in-cylinder values. The engine and operating point specific data can be found in appendix A.2.1 and A.2.2.

In comparison with the non-fired operation, the mean characteristic curve of the fired operation shows a decrease of the ratio l_t/l_f after TDC. Comparable to non-fired operation, the ratio u'/s_l decreases in the same time. Shortly before the 10 % transformation point B, the ratio of l_t/l_f starts to increase until point C (50 % transformation point) while the ratio of u'/s_l decreases further. Reaching the maximal in-cylinder pressure point D, the ratio l_t/l_f finally decreases. At the same time, the ratio u'/s_l is enhanced.

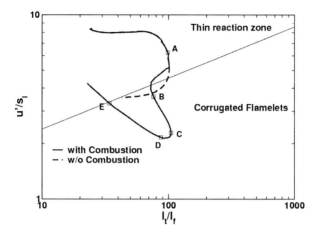

Figure 3.7: Calculated characteristic curves of fired and non-fired highly turbocharged engine at part load operating point (2000 rpm, 4.99 bar) in regime diagram of turbulent premixed combustion. Letters in the figure refer to A: 15 °CA bTDC (spark timing), B: 1 °CA aTDC (x_b=10 %), C: 11 °CA aTDC (x_b=50 %), D: 16 °CA aTDC (p_{max}), E: 29 °CA aTDC (x_b=90 %)

The characteristic curve depicted in figure 3.7 is a mean of the whole in-cylinder characteristic and thus an average of all combustion zones, i.e. the unburnt, the burning and the burnt zone. In order to separate the influence of the different zones on the engine characteristic curve and gain an overview of the complete processes occurring, the distribution of the in-cylinder values from individual grid cells of u'/s_l and l_t/l_f is plotted in the Borghi-Peters diagram classified in unburnt, burning and burnt zone for different crank angle degrees. The zone classification is performed according to:

- *Unburnt zone*: Combustion progress c equals zero (green dots)

- *Burning zone*: In the range $-0.5 l_{f,t} < G < 0.5 l_{f,t}$, according to equation 3.11 (red dots)

- *Burnt zone*: Combustion progress is $0.1 \leq c < 0.5$ (violet dots), combustion progress is $0.5 \leq c < 0.9$ (blue dots), and combustion progress is $0.9 \leq c$ (cyan dots), with c implicitly given by equation 3.12

Figures 3.8 and 3.9 display the combustion zones classified Borghi-Peters diagram for point A to B and point C to E, respectively. Additionally, figures 3.10, 3.11, 3.12 and 3.13 depict the in-cylinder distribution of combustion progress variable, laminar flame speed, gas density, turbulent velocity, integral length scale as well as velocity.

At spark timing (figure 3.8 a) the in-cylinder characteristics distribute over a wide range of length scale ratios $0.1 \leq l_t/l_f \leq 200$ as well as velocity ratios $0.4 \leq u'/s_l \leq 10$ due to a highly non-uniform distribution of the turbulence scales. Apparent from figure 3.11, high values of l_t/l_f and u'/s_l refer to the combustion chamber center, while low values occur close to the wall.

At 7 °CA bTDC (figure 3.8 b) a flame starts to develop and the in-cylinder gas phase close to the spark plug begins to burn. Due to the initial in-cylinder distribution of the turbulence scales, at which close to the spark plug high values of l_t and u' are present, the burning zone appears in the Borghi-Peters diagram first of all in the area of high ratios of l_t/l_f and u'/s_l.

In the following crank angle degrees until point B (figure 3.8 c-f), the flame radius increases continuously and the mixture within the flame is transformed, which is apparent from the increasing combustion progress variable. The turbulent kinetic energy and its dissipation, k and ε, dissipate globally, leading to a decrease of u' and l_t. The unburnt mixture possesses a local increase of l_t/l_f due to a higher dissipation rate of l_t in comparison to l_f, whereas l_f decreases due to an increase of the gas density. The burning mixture is characterised by an increasing flame thickness as a result of decreasing gas density. Thus, the mixture within the flame tends to low ratios of l_t/l_f. This effect dominates the global averaged length scale ratio. Due to an increasing temperature inside the flame, the laminar flame speed is enhanced. As a result the characteristic distribution possessing a reaction progress of $0.1 \leq c < 0.5$ is mainly located in the area of small ratios of u'/s_l. However, as a result of pressure fluctuations, the velocity increases locally within the flame resulting in an increase of u' (and thus u'/s_l) especially in areas possessing a high combustion progress $(0.9 \leq c)$.

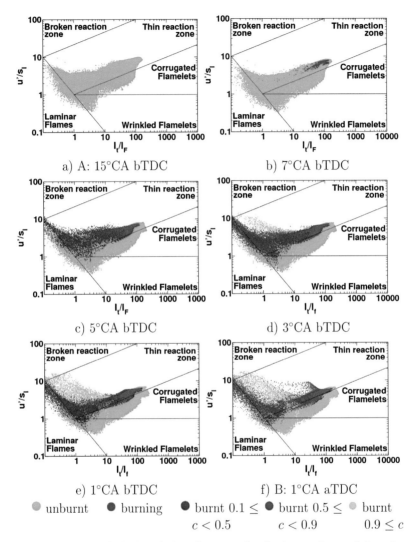

a) A: 15°CA bTDC

b) 7°CA bTDC

c) 5°CA bTDC

d) 3°CA bTDC

e) 1°CA bTDC

f) B: 1°CA aTDC

● unburnt   ● burning   ● burnt $0.1 \leq$ ● burnt $0.5 \leq$ ● burnt
                                              $c < 0.5$ $c < 0.9$ $0.9 \leq c$

Figure 3.8: Calculated distribution of velocity scales and length scales of the fired highly turbocharged engine at full load operating point (1500 rpm, 20.17 bar, tumble flap closed) in regime diagram of turbulent premixed combustion for point A to B in figure 3.7

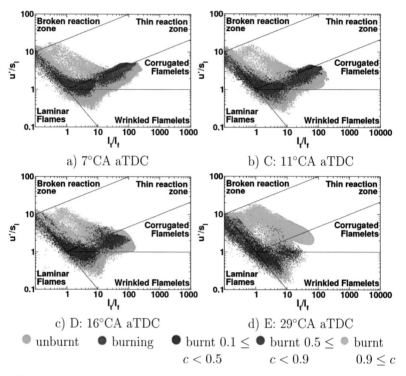

a) 7°CA aTDC   b) C: 11°CA aTDC

c) D: 16°CA aTDC   d) E: 29°CA aTDC

● unburnt   ● burning   ● burnt $0.1 \leq$ ● burnt $0.5 \leq$ ● burnt
$c < 0.5$ $c < 0.9$ $0.9 \leq c$

Figure 3.9: Calculated distribution of velocity scales and length scales of the fired highly turbocharged engine at part load operating point (2000 rpm, 4.99 bar) in regime diagram of turbulent premixed combustion for point C to E in figure 3.7

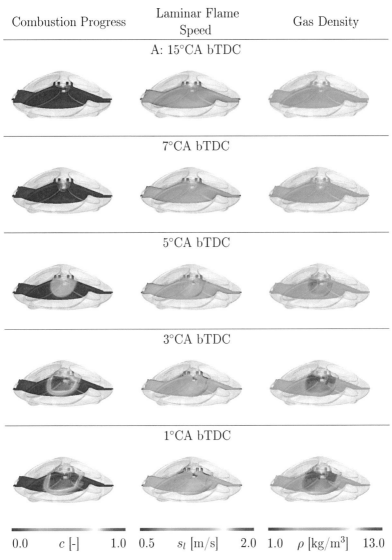

Combustion Progress	Laminar Flame Speed	Gas Density

A: 15°CA bTDC

7°CA bTDC

5°CA bTDC

3°CA bTDC

1°CA bTDC

0.0 c [-]   1.0   0.5 s_l [m/s]   2.0   1.0 ρ [kg/m^3]   13.0

Figure 3.10: Calculated in-cylinder distribution of combustion progress variable, laminar flame speed, and gas density of the fired highly turbocharged engine at part load operating point (2000 rpm, 4.99 bar) for point A to B in figure 3.7

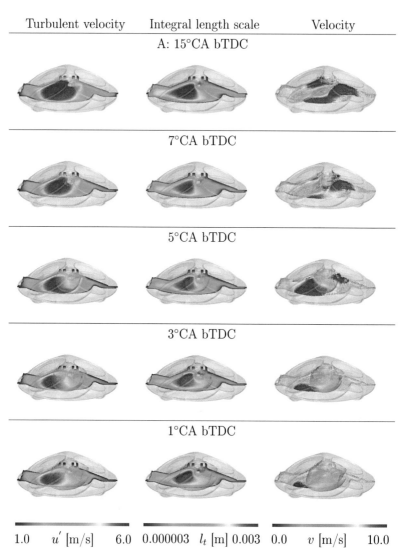

Turbulent velocity	Integral length scale	Velocity

A: 15°CA bTDC

7°CA bTDC

5°CA bTDC

3°CA bTDC

1°CA bTDC

1.0 u' [m/s]   6.0   0.000003 l_t [m] 0.003   0.0 v [m/s]   10.0

Figure 3.11: Calculated in-cylinder distribution of turbulent velocity, integral length scale, and velocity of the fired highly turbocharged engine at part load operating point (2000 rpm, 4.99 bar) for point A to B in figure 3.7

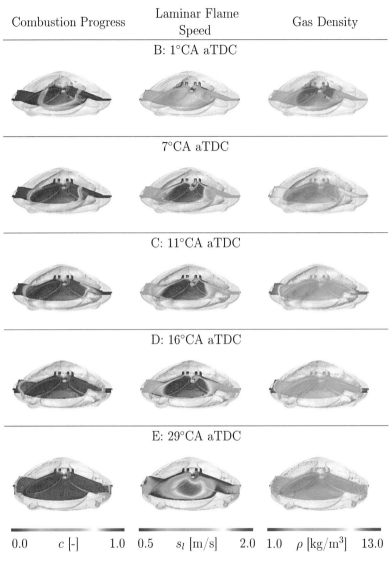

Figure 3.12: Calculated in-cylinder distribution of combustion progress variable, laminar flame speed, and gas density of the fired highly turbocharged engine at part load operating point (2000 rpm, 4.99 bar) for point B to E in figure 3.7

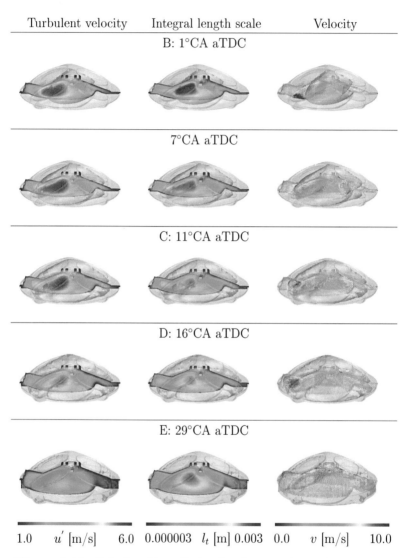

Turbulent velocity	Integral length scale	Velocity

1.0 u' [m/s]   6.0   0.000003 l_t [m]   0.003   0.0 v [m/s]   10.0

Figure 3.13: Calculated in-cylinder distribution of turbulent velocity, integral length scale, and velocity of the fired highly turbocharged engine at part load operating point (2000 rpm, 4.99 bar) for point B to E in figure 3.7

The transformation of the in-cylinder mixture between $x_b = 10\,\%$ (point B) and $x_b = 50\,\%$ (point C) (figures 3.8 f and 3.9 a-b) possesses an increase of the global ratio of l_t/l_f. Both length scales decrease in the unburnt mixture. l_f reduces due to increasing gas density as a result of compression of the unburnt mixture by the propagating flame. l_t decreases due to its dissipation.

The rate of decrease of l_f is greater than the one of l_t, thus the ratio l_t/l_f increases. In the burnt zone the same is true. However, the decrease of l_f is here a result of increasing laminar flame speed (due to temperature increase) as well as increasing gas density. The enhancement of the burnt gas density is a result of the increased pressure gradient between the burnt and unburnt zone close to the wall and resulting mass transfer from the unburnt to the burnt zone. The global ratio of u'/s_l continuous decreasing. The unburnt mixture distribution stays almost constant. However, the velocity ratio of the burnt mixture decreases heavily due to increasing s_l, which becomes apparent from the in-cylinder regions possessing a high combustion progress $0.9 \leq c$.

Between point C ($x_b = 50\,\%$) and the maximal in-cylinder pressure point D (figure 3.9 b-c), where the burning rate is strongly reduced, the global length scale ratio starts to decrease while the velocity scale ratio increases. In the expansion stroke, the global temperature decreases. Following, the laminar flame velocity reduces, which in turn leads to an enhancement of the flame thickness and thus a reduction of l_t/l_f. In the burnt as well as the unburnt zone, the turbulent velocity decreases ongoing due to the dissipation of k. As a result, the laminar flame speed possesses greater values than u' especially in the unburnt zone, thus shifting the unburnt mixture towards lower ratios of u'/s_l.

These processes intensify between point D (maximal in-cylinder pressure) and point E ($x_b = 90\,\%$) (figure 3.9 c-d).

From the combustion zone classified distributions of the length and velocity scales it can be concluded, that the burning of the mixture in a highly turbocharged engine takes place over a wide spread of combustion regimes. Starting at high ratios of u'/s_l in the thin reaction zone regime, the burning zone shifts into the corrugated and finally wrinkled flamelets regime, where the flame extinguishes. The post flame burning process occurs mainly in the thin reaction zone regime. In a later stage of combustion, the post flame zone is shifted into the corrugated and wrinkled flamelet regime. Thus, the flamelet assumption is fulfilled for the entire combustion process of turbocharged direct injection SI engines.

In the following, the results obtained are evaluated in terms of engine application. A change of the initial ratios of u'/s_l and l_t/l_f would lead to an overall shift of the in-cylinder distribution of the velocity and length scales. A positive effect on combustion duration and emission formation can be achieved by utilising the impact of turbulence on laminar flame in the thin reaction zone and corrugated flamelet regime, i.e. towards higher ratios[12] of u'/s_l and l_t/l_f. The higher turbulence scales lead to a flame corrugation and flame stretch, both enhancing the flame propagation and thus decreasing the combustion duration. Additionally, with increasing turbulence level the mixture homogeneity is enhanced, which reduces the emission formation. In contrast, a shift into the wrinkled flamelet regime, i.e. towards lower ratios of u'/s_l and l_t/l_f, has a negative effect on combustion duration and emission formation. The marginal impact of turbulence on laminar flame reduces the flame propagation and increases the combustion duration, which leads to an incomplete mixture transformation and thus emission formation.

3.6   Laminar Flame Speed Closure Formulation

The driving force of flame propagation in the G-equation approach is the turbulent flame speed. This quantity is primarily defined by the turbulence, but also by the fuel-dependent laminar flame speed s_l, which is the velocity of the unburnt gases moving normal and undisturbed to the flame front into the combustion front [96]. The thermo-chemical property is dependent on the fuel-air equivalence ratio ϕ, the temperature in the unburnt mixture T_u, and the pressure p [132].

The exact determination of the laminar flame speed requires the solution of the differential equations for mass, momentum, energy and species mass. Thereby, the species mass conservation equation needs to be solved for all involved species, which leads to several hundreds of differential equations. However, an analytical solution of the differential equation system described is still to be found. A numerical solution would exceed an acceptable time frame of 3D-CFD engine combustion simulation [118]. For this reason, a simple and closed formulation of the

---

[1]In order to prevent an entering of the flame into the broken reaction zone regime and thus flame extinction, it is necessary to simultaneously increase the ratio of l_t/l_f with u'/s_l.

[2]As shown in section 3.5.3, an increase of the ratios u'/s_l and l_t/l_f can be achieved by increasing the integral length scale, increasing the engine revolution speed, and increasing the initial pressure. A decrease of the initial temperature, fuel-air equivalence ratios unequal one, and an increase of the EGR rate lead only to an enhancement of the ratio u'/s_l.

laminar flame speed is required.

In order to estimate the laminar flame speed, different fitting functions are proposed in the literature. These functions are first of all compared to direct numerical calculations. Afterwards, the impact of the laminar flame speed evaluated using the different fitting functions on the turbulent flame speed is investigated in the simplified test case. Thereafter, the formulations are explored in engine test case for varying global fuel-air equivalence ratios.

3.6.1 Direct Numerical Calculation of Laminar Flame Speed

In order to calculate the laminar flame speed numerically, we need to consider a planar steady state flame normal to the x-direction with the unburnt mixture at x $->$ $-\infty$ and the burnt mixture at x $->$ $+\infty$. Assuming a constant mass flux through the planar steady state flame, the one-dimensional balance equations for continuity, mass fraction of chemical species i and energy read [132]

$$\frac{\partial (\rho v)}{\partial x} = 0 \tag{3.21}$$

$$\rho v \frac{\partial Y_i}{\partial x} = -\frac{\partial j_i}{\partial x} + \omega_i \tag{3.22}$$

$$c_p \rho v \frac{\partial T}{\partial x} = \frac{\partial}{\partial x}\left(\lambda \frac{\partial T}{\partial x}\right) - \sum_{i=1}^{N} h_i \omega_i - \sum_{i=1}^{N} c_{p,i} j_i \frac{\partial T}{\partial x} + q_{rad}. \tag{3.23}$$

Herein, v is the velocity in x-direction, λ is the thermal conductivity, j_i is the diffusion flux of species i, h_i is the specific enthalpy of species i, c_p and $c_{p,i}$ are the specific heat capacity of mixture and species i, respectively, and N denotes the total number of species. The chemical source term ω_i represents the mass of species i produced

$$\omega_i = M_{W,i} \sum_{k=1}^{R} \nu_{k,i} w_k \tag{3.24}$$

where $M_{W,i}$ is the molecular weight of species i, $\nu_{k,i}$ is the stoichiometric coefficient of species i in reaction k, and w_k is the rate of reaction k in a mechanism containing R chemical reactions

$$w_k = k_{fk}(T) \prod_{i=1}^{N} \left(\frac{\rho Y_i}{M_{W,i}}\right)^{\nu'_{k,i}} - k_{bk}(T) \prod_{i=1}^{N} \left(\frac{\rho Y_i}{M_{W,i}}\right)^{\nu''_{k,i}}. \tag{3.25}$$

The stoichiometric coefficients of the forward and backward step for species i in reaction k are denoted by $\nu'_{k,i}$ and $\nu''_{k,i}$. The rate coefficients $k_{fk}(T)$ and $k_{bk}(T)$ are expressed in the form [60]

$$k_k = A_k \left(\frac{T}{T_0}\right)^n \exp\left(\frac{-E_a}{RT}\right). \tag{3.26}$$

Integrating the continuity equation defines the laminar flame speed s_l as an eigenvalue [60]

$$\rho v = \rho_u s_l \tag{3.27}$$

which needs to be varied in the numerical calculation, until a steady-state solution is obtained [136].

3.6.2 Laminar Flame Speed Fitting Functions

Different fitting functions of the laminar flame speed proposed in the literature are introduced in the following. The formulations presented primarily distinguish either in theoretical considerations (Mallard and LeChatelier, Perlman) or in empirical studies (Metghalchi and Keck, Gülder).

Mallard and LeChatelier

Mallard and LeChatelier's formulation bases on theoretical consideration of the laminar flame structure [96]. They assume the temperature as the driving force of s_l, and derived the formulation

$$s_l \propto \sqrt{\exp\left(\frac{-E_a}{RT}\right)} = \exp\left(\frac{-E_a}{2RT}\right). \tag{3.28}$$

Taking into account that most of chemical reaction takes place at temperatures close to the adiabatic flame temperature T_b, the temperature T in equation 3.28 can be defined as $T = T_b - T_f$ [150], yielding

$$s_l = s_{l,max} \exp\left(\frac{E_a}{2R}\left(\frac{1}{T_b} - \frac{1}{T_f}\right)\right). \tag{3.29}$$

Metghalchi and Keck

Metghalchi and Keck [119] derived their formulation by utilising laminar flame speed measurement data. The resulting formulation describes the laminar flame speed as a function of fuel-air equivalence ratio ϕ, temperature T, pressure p, and diluent fraction ψ, weighted by empirical determined fuel specific constants, which are listed in table 3.3.

$$s_l = B_m + B_\phi \left(\phi - \phi_m\right)^2 \left(\frac{T}{T_0}\right)^\alpha \left(\frac{p}{p_0}\right)^\beta (1 - 2.1\psi) \qquad (3.30)$$

B_m	B_ϕ	ϕ_m	α	β
[cm/s]	[cm/s]	[-]	[-]	[-]
26.32	-84.72	1.13	$2.18 - 0.84(\phi - 1)$	$-0.16 + 0.22(\phi - 1)$

Table 3.3: Fuel-specific constants determined using equation 3.30 for *iso*-Octane [119]

The formulation is validated over the range $0.8 \leq \phi \leq 1.4$, $300 \leq T \leq 700$ K, $0.413 \leq p \leq 50.013$ bar, and $0 \leq \psi \leq 0.2$.

Gülder

Gülder [61] describes the laminar flame speed using an empirical formula, too. His formulation reads

$$s_l = ZW\phi^\eta \exp\left(-\xi\left(\phi - 1.075\right)^2\right) \left(\frac{T}{T_0}\right)^\alpha \left(\frac{p}{p_0}\right)^\beta (1.0 - 2.3\psi).$$
$$(3.31)$$

The corresponding fitting coefficients are listed in table 3.4.

Z	W	η	ξ	α	β
[-]	[m/s]	η[-]	[-]	[-]	[-]
1	0.4658	-0.326	4.48	1.56	-0.22

Table 3.4: Fuel-specific constants determined using equation 3.31 for *iso*-Octane [61]

The formulation is validated for $300 \leq T \leq 700$ K, $1 \leq p \leq 40$ bar, and $\psi = 0$.

Perlman

The formulation of Perlman [131] is derived on the basis of asymptotic analysis.

Following Göttgens et al. [60], flames are assumed to have an inner layer, with a characteristic temperature T_δ called inner layer temperature, in which reaction takes place. The inner layer temperature is assumed to be a function of pressure only [136]:

$$T_\delta = -\frac{E}{\ln{(p/B)}} \qquad (3.32)$$

with B and E as modelling constants. The adiabatic flame temperature T_b can be calculated via [131]

$$T_b = C_0 + C_1\phi + C_2\phi^2 + C_3\phi^3 + C_4T_u + C_5T_u\psi + C_6\log{(p)} \qquad (3.33)$$

where T_u is the temperature of the surroundings at the beginning of the calculation, and C_1 to C_6 are fuel-dependent constants. By analysing fuel lean conditions, Göttgens et al. [60] derived a fitting function for the laminar flame speed reading

$$s_l = F\,Y_{F,u}^m\,\exp{\left(-\frac{G}{T_\delta}\right)}\frac{T_u}{T_\delta}\left(\frac{T_b - T_\delta}{T_b - T_u}\right)^n. \qquad (3.34)$$

Herein, the flame speed depends on the fuel-air equivalence ratio through the mass fraction of the fuel in the unburnt mixture $Y_{F,u}$. The fuel dependent coefficients B, E, F, and G as well as n and m need to be adjusted in order to match numerical data.

To extend the formulation of Göttgens et al. for fuel rich conditions, Perlman [131] introduced a term representing the mass fraction of unburnt oxygen at the beginning of calculation $Y_{O_2}^l$, by transferring the fuel-air equivalence ratio ϕ space into the mixture fraction Z space

$$\phi = \nu_k\frac{Y_{F,u}}{Y_{O_2,u}} = \frac{Z}{1-Z}\frac{(1-Z_{st})}{Z_{st}}, \qquad (3.35)$$

yielding

$$s_l = F\,Y_{O_2}^l\,\exp{\left(-\frac{G}{T_\delta}\right)}\left(\frac{T_u}{T_\delta}\right)^k\left(\frac{T_b - T_\delta}{T_b - T_u}\right)^n Z^q\,(1-Z)^w. \qquad (3.36)$$

The fuel dependent approximation coefficients l, k, q, and w need to be adjusted, too.

For inner layer temperatures and adiabatic temperatures, fitting is done to all data at once whereat the flame speed is fitted using interpolation both in equivalence ratio and temperature space. A linear interpolation is performed every 100 K step and for ϕ interpolation occurs for $0.7 \leq \phi \leq 1.0, 1.1 < \phi < 1.5$, as well as $2.1 \leq \phi \leq 2.3$. To ensure smoothness a second degree Lagrange polynomial is computed for every equivalence ratio point requiring interpolation, using points on both sides of the interval in question as interpolation data [131]. The fitting is done for an iso-Octane / n-Heptane reaction mechanisms of Ahmed et al. [6] in the range $0.2 \leq \phi \leq 4.0$, $T \geq 450$ K, $1 \leq p \leq 130$ bar, and $0 \leq \psi \leq 0.5$, where EGR is assumed to consists of products of complete combustion, i.e. CO_2, H_2O and N_2.

3.6.3 Comparison between Laminar Flame Speed Fitting Functions and Direct Numerical Calculations

In the following, the laminar flame speeds obtained using the different fitting functions are compared with direct numerical calculations. This comparison implies the validation of the reaction mechanism used for direct numerical calculation. Therefore, the detailed numerical calculation of the laminar flame speed using an iso-Octane / n-Heptane reaction mechanism of Ahmed et al. [6] is compared to measurement data, first.

Measurements of laminar burning velocities of premixed hydrocarbon-air flames are limited to low pressures and temperatures, where the fuel-air charge can be controlled allowing measurements at precisely defined values [74]. Depending on the measurement technique, the laminar flame speed determined may vary from author to author. A detailed analysis of the differences in burning velocities obtained using different measurement technique can be found in [74].

Figure 3.14 displays the calculated and measured laminar flame speeds of iso-Octane as a function of fuel-air equivalence ratio in dependency of pressure and temperature. The measurement data presented are obtained by Jerzembeck et al. [75], Heimel et al. [66], and Davis et al. [41].

These plots reveal the overall dependency of the laminar flame speed on fuel-air equivalence ratio, temperature and pressure. The laminar flame speed of iso-Octane peaks at a fuel-air equivalence ratio close to $\phi = 1.1$ and decreases with ϕ values greater and smaller than the peak value. An increase of pressure results in a decrease of the laminar flame speed. An increase of the temperature leads to an increase of the laminar flame speed.

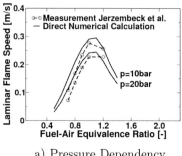 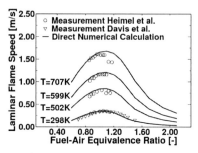

a) Pressure Dependency   b) Temperature Dependency

Figure 3.14: Measured [41, 66, 75] and direct numerically calculated laminar flame speed velocities of *iso*-Octane as a function of fuel-air equivalence ratio and a) different pressures at $T = 373$ K and b) different temperatures at $p = 1$ bar

Basically, the direct numerically calculated flame speeds match the overall trend of measurement data well. However, an inconsistence between calculated and measured flame speed velocities can be observed primarily for fuel-air equivalence ratios smaller than 1.0 at 373 K and a pressure of 10 and 20 bar. In this range, the Mean Absolute Percentage Error (MAPE) is 19.93 % and 25.29 %, while in the full ϕ range MAPE is 16.21 % and 18.9 %, respectively. Concerning the temperature reproduction at $p = 1$ bar, the MAPE is 9.01 % at $T = 707$ K, 3.65 % at $T = 599$ K, 8.27 % at $T = 502$ K, and 0.53 % at $T = 298$ K. In general, the MAPE decreases with decreasing pressure and temperature values, which can be assigned to increasing difficulties in charge controlling arising in measurements with increasing temperature and pressure values.

In order to evaluate the different fitting functions of the laminar flame speed for engine-like conditions, the flame speeds obtained using the formulations of Mallard and LeChatelier, Metghalchi and Keck, Gülder as well as Perlman are compared to direct numerically calculated velocities in the following.

Figure 3.15 displays the numerically calculated flame speeds for different fuel-air equivalence ratios and pressures at a temperature of 750 K. Furthermore, the figure represents the laminar flame speeds obtained using the different fitting formulations.

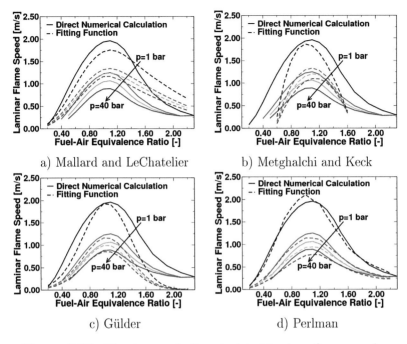

a) Mallard and LeChatelier   b) Metghalchi and Keck

c) Gülder   d) Perlman

Figure 3.15: Direct numerically calculated laminar flame speed velocities vs. velocities obtained using different fitting functions as a function of fuel-air equivalence ratio at $p = 1, 10, 20, 40$ bar and $T = 750$ K

Considering the fitting function of Mallard and LeChatelier (figure 3.15 a), the difference of the burning velocity between fitting function and direct numerically calculation increases for pressures greater and smaller than 10 bar. In detail, in the medium range of $0.8 \leq \phi \leq 1.2$ the MAPEs are 11.56 % at $p = 1$ bar, 8.75 % at $p = 10$ bar, 18.94 % at $p = 20$ bar, and 33.16 % at $p = 40$ bar. With a MAPE of 28.32 % for all pressures investigated, the formulation depicts the laminar flame speed on the fuel lean side ($\phi \leq 1$) roughly. Additionally, using the fitting function the burning velocity decreases almost linearly with increasing fuel-air equivalence ratio on the fuel rich side ($\phi \geq 1$), which is not in line with the direct calculated velocities. The resulting MAPE in this range is 39.82 %.

The formulation of Metghalchi and Keck (figure 3.15 b) shows in the medium range of $0.8 \leq \phi \leq 1.2$ a MAPE of $7.49\,\%$ at $p = 1$ bar, $3.28\,\%$ at $p = 10$ bar, $9.41\,\%$ at $p = 20$ bar, and $17.66\,\%$ at $p = 40$ bar. Thus, as observable for the formulation of Mallard and LeChatelier, differences between the fitting function and the direct numerical calculated flame speed data increase for pressures greater and smaller than 10 bar. Furthermore, using the formulation of Metghalchi and Keck, the fuel-air equivalence ratio value possessing the maximum burning velocity shifts towards higher values with increasing pressure. This effect is neither visible from the results obtained by direct numerical calculation nor from the measurement data represented in figure 3.14. Considering the reproduction of the decreasing laminar flame speed on the fuel lean and fuel rich side, the fitting function underestimates the burning velocities for fuel-air equivalence ratios smaller than 0.8 and greater than 1.4 by far, leading to a narrow combustible range. Latter values constitute the range the formulation is validated for. In this range the resulting MAPEs are $5.93\,\%$ on the fuel lean side $(0.8 \leq \phi \leq 1.0)$ and $17.38\,\%$ on the fuel rich side $(1.0 \leq \phi \leq 1.4)$.

Using the fitting function of Gülder (figure 3.15 c), the maximum burning velocities match the velocities obtained by direct numerical calculation quite well. In the medium range of $0.8 \leq \phi \leq 1.2$ the MAPE are $5.01\,\%$ at $p = 1$ bar, $7.68\,\%$ at $p = 10$ bar, $6.26\,\%$ at $p = 20$ bar, and $3.46\,\%$ at $p = 40$ bar. On the fuel lean side $(\phi \leq 1)$, the laminar flame speed is slightly underestimated with a MAPE of $7.31\,\%$. On the fuel rich side $(\phi \geq 1)$, the formulation underestimates the laminar flame velocity by far, limiting the combustible range to fuel-air equivalence ratios smaller than 1.8. The resulting MAPE is $38.28\,\%$.

Regarding the fitting formulation of Perlman (figure 3.15 d), the MAPEs are in the medium range of $0.8 \leq \phi \leq 1.2$ $5.41\,\%$ at $p = 1$ bar, $7.33\,\%$ at $p = 10$ bar, $10.29\,\%$ at $p = 20$ bar, and $12.74\,\%$ at $p = 40$ bar. Thus, with increasing pressure the difference between direct numerically calculated flame speed and flame speed obtained using the fitting function increases. However, the split fitting procedure leads to a good reproduction of the laminar burning velocities on the fuel lean as well as fuel rich side. Here, the MAPE are $3.77\,\%$ $(\phi \leq 1)$ and $3.91\,\%$ $(\phi \geq 1)$, respectively.

Figure 3.16 compares the direct numerically calculated laminar flame speeds to the fitting functions for different fuel-air equivalence ratios and temperatures at $p = 20$ bar.

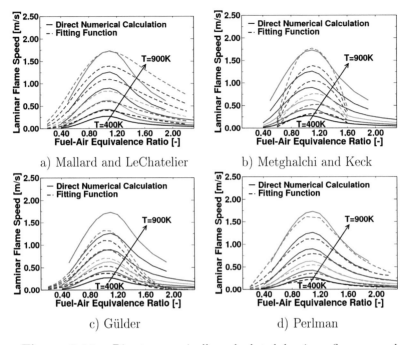

Figure 3.16: Direct numerically calculated laminar flame speed velocities vs. velocities obtained using different fitting functions as a function of fuel-air equivalence ratio at $T = 400, 500, 600, 700, 800, 900$ K and $p = 20$ bar

Using the fitting formulation of Mallard and LeChatelier (figure 3.16 a), the difference to the direct numerically calculated flame speeds decreases with increasing temperature. In the medium range of $0.8 \leq \phi \leq 1.2$ the MAPEs are 52.21 % at $T = 400$ K, 46.47 % at $T = 500$ K, 36.4 % at $T = 600$ K, 25.02 % at $T = 700$ K, 12.85 % at $T = 800$ K, and 1.06 % at $T = 900$ K. On the fuel lean side ($\phi \leq 1$), the formulation overestimates the burning velocity by far with a MAPE of 60.15 %. The same is true on the fuel rich side ($\phi \geq 1$) with a MAPE of 61.34 %.

The formulation of Metghalchi and Keck (figure 3.16 b) matches the maximum burning velocity for $800 \leq T \leq 900$ K quite well. However, like the formulation of Mallard and LeChatelier, the MAPE increases with decreasing temperature. In detail, in the medium range of $0.8 \leq \phi \leq 1.2$ the MAPEs are 13.4 % at $T = 400$ K, 17.35 % at

$T = 500$ K, 16.48% at $T = 600$ K, 12.21% at $T = 700$ K, 6.0% at $T = 800$ K, and 1.43% at $T = 900$ K. Besides, the shifting effect of the maximum burning velocity from fuel lean to fuel rich side appears for decreasing temperatures, too. In line with the results obtained by varying the pressure, the laminar flame speed is underestimated by far for fuel-air equivalence ratios smaller 0.8 and greater 1.4, leading to a narrow combustible range. In the range the formulation is validated for the MAPEs are 6.16% on the fuel lean side ($0.8 \leq \phi \leq 1.0$) and 25.82% on the fuel rich side ($1.0 \leq \phi \leq 1.4$).

The Gülder formulation (figure 3.16 c) matches the laminar flame speed in the medium range $0.8 \leq \phi \leq 1.2$ best for a temperature of 700 K. For temperatures smaller and greater than 700 K, the formulation overestimates and underestimates the flame speed, respectively. The MAPEs are here 42.77% at $T = 400$ K, 28.94% at $T = 500$ K, 14.47% at $T = 600$ K, 0.31% at $T = 700$ K, 12.73% at $T = 800$ K, and 24.53% at $T = 900$ K. Regarding the velocities predicted on the fuel lean side ($\phi \leq 1$), the fitting function matches the direct numerically calculated values roughly with a MAPE of 50.79%. On the fuel rich side ($\phi \geq 1$) the burning velocities are underestimated increasingly with ϕ, leading to burning velocities close to zero for fuel-air equivalence ratios greater 1.8. The MAPE is here 25.01%.

Using the fitting function of Perlman (figure 3.16 d), the maximum flame speed is in the medium range of $0.8 \leq \phi \leq 1.2$ increasingly under-estimated with decreasing temperatures compared to direct numerically calculated velocities. The MAPEs are here 17.53% at $T = 400$ K, 15.6% at $T = 500$ K, 13.36% at $T = 600$ K, 11.55% at $T = 700$ K, 9.81% at $T = 800$ K, and 5.55% at $T = 900$ K. Nevertheless, the flame speed is well reproduced on the fuel lean as well as fuel rich side due to the split fitting procedure. The MAPE are here 2.73% ($\phi \leq 1$) and 6.65% ($\phi \geq 1$), respectively.

Considering the full pressure and temperature range investigated, the formulation of Metghalchi and Keck as well as the formulation of Perlman predict the laminar flame speed in the medium range of $0.8 \leq \phi \leq 1.2$ best. On the fuel lean and fuel rich side, the formula-tion of Perlman possesses the smallest error in representing the direct numerically calculated laminar flame speed.

However, the turbulent flame propagation (and consequently the heat release) is rather defined by the turbulence than by the laminar flame speed [70]. The necessity of a good fitting quality for maximum flame speeds as well as flame speeds under extremely fuel lean and fuel rich conditions is thus an open question, which is addressed in the following sections.

3.6.4 Influence of Laminar Flame Speed Fitting Function on Pressure Trace in Simplified Test Case

In order to investigate the impact of the laminar flame speed on the turbulent flame speed and consequently the pressure raise in the range of maximum burning velocity in case of a freely propagating turbulent flame, the simplified test case specified in appendix A.1.2 is used in the following. The adiabatic pressure vessel is homogeneously initialised with engine-like conditions at spark timing, i.e. at a temperature of 800 K and a pressure of 10 bar. The initial turbulent velocity is set to 2.5 m/s and the fuel-air equivalence ratio to 0.8, 1.0 and 1.2. For numerical investigation, the G-equation model is used in connection with the simplified closure formulation [17, 31, 112, 170] of the turbulent flame speed

$$\frac{s_t}{s_l} = 1 + \frac{u^{'}}{s_l}. \qquad (3.37)$$

Figures 3.17, 3.18, and 3.19 depict the evolution over time of global values for a) laminar flame speed, b) resultant turbulent flame speed and c) total pressure in the simplified test case for three fuel-air equivalence ratios by using different laminar flame speed fitting functions.

Generally, s_l increases over time in the unburnt mixture in front of the propagating flame as a result of increasing temperature due to flame progress, apparent from pressure raise. In contrary, the turbulent velocity decreases. Since $u^{'} > s_l$, the resulting turbulent flame speed decreases as well.

For $\phi = 0.8$ the initial laminar flame speeds obtained using the different fitting functions vary in the range of 1.0 (formulation of Gülder) to 1.23 m/s (formulation of Mallard and LeChatelier). The ranking of the different fitting functions is in line with the laminar flame speeds represented in figure 3.15. Since $u^{'}$ decreases in case of a freely propagating flame identically for all cases investigated, the different formulations of s_l affect s_t considerably. The resulting initial s_t values range between 3.35 (formulation of Gülder) and 3.58 m/s (formulation of Mallard and LeChatelier). The impact of the initial s_l becomes much more apparent at the end of calculation, at $t = 0.006$ s. Using Gülder's formulation, the resulting in-cylinder pressure is much lower compared to the other formulations due to less mass transformation as a result of low laminar flame speed. The pressure obtained using the laminar flame speed for-

mulation of Mallard and LeChatelier is almost 3 bar higher compared
to the values obtained using Gülder's formulation.

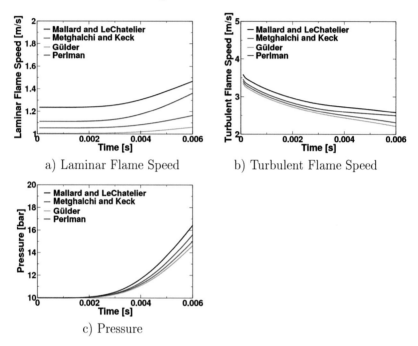

a) Laminar Flame Speed   b) Turbulent Flame Speed

c) Pressure

Figure 3.17: Influence of different laminar flame speed fitting functions on turbulent flame speed and pressure calculated in the simplified
test case initialised with $\phi = 0.8$

The impact of the different laminar flame speed fitting functions on
the turbulent flame speeds and resulting pressure raises consolidate and
become much more obvious for $\phi = 1.0$ and 1.2.

For $\phi = 1.0$, the initial s_l ranges from 1.28 (formulation of Gülder)
to 1.47 m/s (formulation of Mallard and LeChatelier). The resulting
initial s_t values cover the range 3.63 to 3.82 m/s and as consequence the
pressure values range 24.7 to 27 bar at $t = 0.006$ s.

For $\phi = 1.2$ the initial difference in s_l is even greater than for $\phi =$
1.0, i.e. 0.32 m/s. Here, the shifting of the maximal burning velocity
from fuel lean to fuel rich side of the fitting function of Metghalchi
and Keck (as already described in section 3.6.3) becomes obvious. In
comparison to $\phi = 1.0$, the laminar flame speed predicted at $\phi = 1.2$

is higher. The other fitting functions show a comparable or even lower value of s_l than for $\phi = 1$. As a result of different initial s_l values, the initial turbulent flame speed ranges between 3.5 and 3.82 m/s and the pressure at $t = 0.006$ s between 23.8 and 27.4 bar.

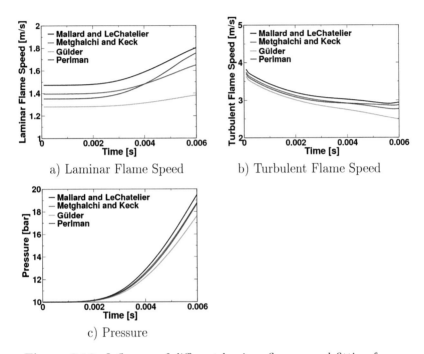

a) Laminar Flame Speed   b) Turbulent Flame Speed

c) Pressure

Figure 3.18: Influence of different laminar flame speed fitting functions on turbulent flame speed and pressure calculated in the simplified test case initialised with $\phi = 1.0$

Summarising, in the range of maximum burning velocity the different fitting functions of the laminar flame speed impact the pressure raise resulting from a freely propagating turbulent flame. Thereby, the maximum pressures predicted differ by 20.55 % at $\phi = 0.8$, 13.14 % at $\phi = 1.0$, and 21.82 % at $\phi = 1.2$.

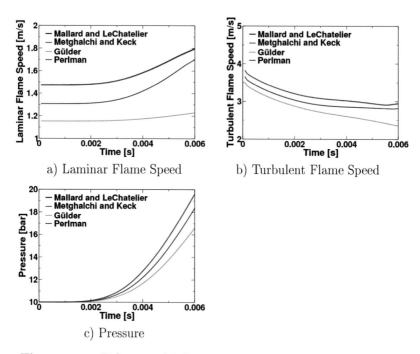

a) Laminar Flame Speed   b) Turbulent Flame Speed

c) Pressure

Figure 3.19: Influence of different laminar flame speed fitting functions on turbulent flame speed and pressure calculated in the simplified test case initialised with $\phi = 1.2$

3.6.5   Influence of Laminar Flame Speed Fitting Function in Engine Test Case

The impact of the laminar flame speed on the pressure trace in engine test case is investigated in the following. Moreover, the necessity of a good fitting quality for flame speeds under extremely fuel lean and fuel rich conditions is addressed.

The different fitting functions of the laminar flame speed are investigated on a part load operating point at 2000 rpm and 4.99 bar of the highly turbocharged SI engine for varying fuel-air equivalence ratios, specified in appendix A.2.1 and A.2.2. In test bed measurements, three different global fuel-air equivalence ratios were explored, i.e. $\phi = 0.9$, 1.0 and 1.3. The inhomogeneity of in-cylinder ϕ values are depicted in figure 3.20 as a function of turbulent velocity at spark timing.

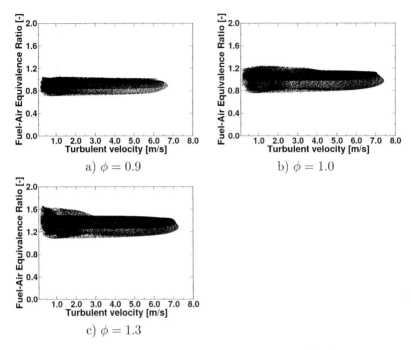

a) $\phi = 0.9$ b) $\phi = 1.0$

c) $\phi = 1.3$

Figure 3.20: Calculated in-cylinder distribution of fuel-air equivalence ratio as a function of turbulent velocity at spark timing of the highly turbocharged SI engine at part load operating point (2000 rpm, 4.99 bar) for varying global fuel-air equivalence ratios

For a global $\phi = 0.9$ the local in-cylinder fuel-air equivalence ratio varies between $0.7 \leq \phi \leq 1.1$. The stoichiometric point $\phi = 1.0$ shows a variation between $0.72 \leq \phi \leq 1.3$. A further enrichment ($\phi = 1.3$) leads to a wider spread of fuel-air equivalence ratio between $\phi = 1.05$ and 1.7.

In order to compare the different approaches in numerical calculation, the G-equation model is used with the simplified closure formulation [2, 81] of the turbulent flame speed

$$\frac{s_t}{s_l} = 1 + C \left(\frac{u^{'}}{s_l}\right)^{0.7} \qquad (3.38)$$

where C is a tuning coefficient, which needs to be fitted. The fitting is done using the Perlman formulation of the laminar flame speed. The resulting C value is kept constant for all calculations.

Figure 3.21 displays the pressure curves obtained using the different fitting functions together with the measured in-cylinder pressures.

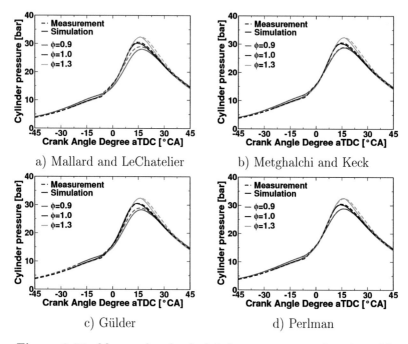

a) Mallard and LeChatelier   b) Metghalchi and Keck

c) Gülder   d) Perlman

Figure 3.21: Measured and calculated pressure curves based on different s_l fitting functions of highly turbocharged engine at part load operating point (2000 rpm, 4.99 bar) for varying fuel-air equivalence ratios

Using the fitting function of Mallard and LeChatelier (figure 3.21 a), the measured pressure curve is well reproduced for fuel rich conditions. For fuel lean conditions, the in-cylinder pressure is underestimated by 2.5 bar. For a global $\phi = 1.0$ the formulation slightly underestimates the measured pressure curve. The formulation of Metghalchi and Keck (figure 3.21 b) leads to a good reproduction of the measured pressure curves for all global fuel-air equivalence ratios investigated. Only for stoichiometric conditions, a small underestimation of the measured pressure can be observed. Gülder's fitting function (figure 3.21 c) leads to a well matching of the pressure curve for fuel-air equivalence ratio of 1.3. For stoichiometric conditions, the formulation slightly overestimates the in-cylinder pressure. For $\phi = 0.9$ the in-cylinder pressure is underestimated by 1.5 bar. The formulation of Perlman leads to a good reproduction of the measured in-cylinder pressure curve at $\phi = 0.9$. For $\phi = 1.0$ the formulation slightly underestimates the in-cylinder pressure, while for $\phi = 1.3$ the in-cylinder pressure is slightly overestimated.

In summary, the fitting functions predict comparable pressure curves for fuel rich conditions. Note that all fitting functions reproduce the maximum measured pressure well. However, all formulations underestimate the measured in-cylinder pressure after 15 °CA aTDC for $\phi = 1.3$. Under stoichiometric conditions, small differences in calculated in-cylinder pressures can be observed. The differences in maximum cylinder pressure increase up to 2.5 bar for fuel lean conditions. The formulation of Mallard and LeChatelier as well as the formulation of Gülder possess the biggest differences compared with measurement data, while the formulations of Methgalchi and Keck as well as Perlman match the measurement data well. Latter outcome is in line with the results obtained from the comparison with direct numerically calculated flame speeds in section 3.6.3.

With a difference of in-cylinder pressure prediction of approximately 10 %, the laminar flame speed has a great impact on pressure trace in engine test case. Nevertheless, compared with the simplified test case, the laminar flame speed has less impact on pressure trace in engine test case. This can be assigned to the higher turbulent velocity present in engine test case (see figure 3.20) compared to the one specified in the simplified test case investigation. However, as shown in section 3.5.4, the in-cylinder turbulence level decreases strongly in the expansion stroke, which indicates an increasing impact of laminar flame speed on flame propagation. For detailed investigation, the magnitudes of turbulent velocity and laminar flame speed are displayed exemplarily in figure 3.22 for different crank angles for the case $\phi = 1.0$. Here, Perlman's fitting function for laminar flame speed is used.

Figure 3.23 displays the corresponding distribution of unburnt fuel mass in the cylinder.

At TDC the turbulent velocity possesses higher values than the laminar flame speed, thus being the driving force of the turbulent flame propagation. 15 °CA later the turbulence intensity decreases strongly and the laminar flame speed enhances due to temperature increase in the unburnt mixture.

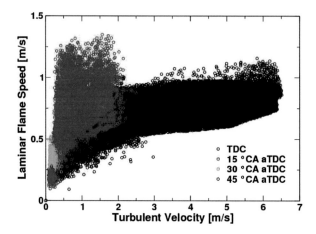

Figure 3.22: Calculated magnitudes of turbulent velocity and laminar flame speed in the unburnt mixture at different crank angle degrees of the highly turbocharged engine at part load operating point (2000 rpm, 4.99 bar) for $\phi = 1.0$

| 15 °CA aTDC | 30 °CA aTDC | 45 °CA aTDC |

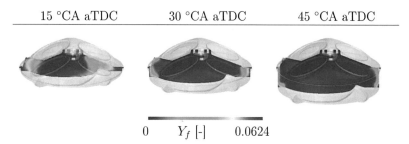

0   Y_f [-]   0.0624

Figure 3.23: Calculated in-cylinder distribution of unburnt fuel mass fraction of highly turbocharged engine at part load operating point (2000 rpm, 4.99 bar) for $\phi = 1.0$

At 30 °CA aTDC more than 90 % of the in-cylinder mass is transformed, the flame is close to the wall where the turbulent velocity is further reduced. Since the laminar flame speed is in the same order of magnitude as the turbulent velocity or even higher close to the wall, its effect on the consumption rate becomes predominant. The associated heat release rate of consumption of the remaining 10 % does not affect the in-cylinder pressure. However, the consumption of the remaining cylinder charge plays an important role in HC emission formation [118, 137]. Therefore, the unburnt fuel mass is used as criterion to evaluate the impact of the different formulations of the laminar flame speed in the following.

Figure 3.24 depicts the global fuel mass fractions calculated using the different fitting functions, normalised with the fuel mass fractions obtained with Perlman's formulation, at 15, 30 and 45 °CA aTDC.

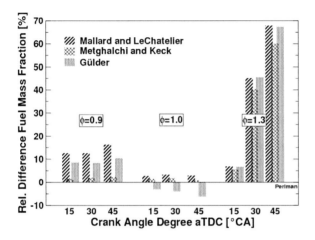

Figure 3.24: Calculated unburnt fuel mass fractions at different crank angle degrees of highly turbocharged engine at part load operating point (2000 rpm, 4.99 bar) for varying fuel-air equivalence ratios using different laminar flame speed fitting functions. Mass fractions are normalised with fuel mass fraction obtained with Perlman's formulation.

Under fuel lean conditions, the fuel mass fraction still present in the combustion chamber at 15 °CA aTDC varies in a range of 12 % using the different fitting functions. The range increases up to 16 % at 45 °CA aTDC. The fuel mass fraction values are an inverse of the maximum in-cylinder pressure values calculated using the different fitting functions for all crank angle degrees investigated, i.e. a decrease in maximum peak pressure predicted results in an increase of unburnt fuel. Following, the unburnt fuel calculated is directly linked to predicted combustion rate.

Considering stoichiometric conditions (figure 3.24 b), the unburnt fuel mass predicted using the different fitting functions varies in a small magnitude of 1 % at 15 °CA aTDC. The difference in unburnt fuel mass increases up to 6 % at 45 °CA aTDC. The unburnt fuel mass predicted is an inverse of maximum in-cylinder pressure here, too.

For $\phi = 1.3$ (figure 3.24 c) the difference at 15 °CA aTDC of unburnt fuel mass is in the range of 7 %. The difference increases strongly up to 45 % at 30 °CA aTDC. A further increase can be seen at 45 °CA aTDC up to 68 %. However, the differences can not be deviated from in-cylinder pressure predictions since the calculated pressures match each other.

Summarising, the difference in unburnt fuel mass prediction in the expansion stroke using the different fitting functions is at 6 % relatively low for stoichiometric conditions. For fuel lean conditions, the difference increases up to 16 %. The greatest difference can be observed for $\phi = 1.3$ with 68 %. The differences of the unburnt fuel mass fraction for $\phi = 0.9$ and 1.0 can be explained with the differences in cylinder pressure predictions of the fitting functions. For $\phi = 1.3$ the pressure curves match each other, and the different fuel mass fractions detected can only be ascribed to the increasing impact of the laminar flame speed on the turbulent flame propagation in the cylinder expansion stroke. In the spread of fuel-air equivalence ratio space present at $\phi = 1.3$, i.e. $1.05 \leq \phi \leq 1.7$, the different fitting functions possess the greatest differences of laminar flame speed representation, which explains the large differences in predicted fuel mass.

Thus, for correct emission formation prediction under engine-relevant conditions, it is necessary that a fitting function predicts the maximum burning rate as well as the flame speeds under extremely fuel lean and fuel rich conditions correctly.

3.7   Turbulent Flame Speed Closure Formulation

In order to close the G-equation model formulation, an expression of the turbulent burning velocity s_t is required. As outlined in section 3.5, premixed combustion takes place primarily in wrinkled flamelets, corrugated flamelets and thin reaction zone regimes. In latter two regimes, the flame front is affected by the turbulent flow. Therefore, an adequate formulation of the turbulent flame velocity must be found describing the impact of chemistry and turbulence on flame propagation in the regimes of turbulent combustion.

For both regimes, Damköhler [40] proposed formulations for the turbulent burning velocity. In his investigations, he considered regimes of small scale and large scale turbulence, which correspond to the thin reaction zone regime and the corrugated flamelets regime. For large scale, small intensity turbulence Damköhler assumed a purely kinematic interaction between a wrinkled flame front and the turbulent flow field [134]

$$\frac{s_t}{s_l} = 1 + \frac{u'}{s_l} \tag{3.39}$$

which leads at the limit $u' \gg s_l$ to

$$s_t \propto u' \tag{3.40}$$

describing a turbulent flame front propagation only due to turbulent fluctuations and independently of laminar velocity, length and time scales [135]. For small scale, high intensity turbulence, Damköhler found that the flame can be altered additionally through viscosity and characteristic length scale l_t [56].

$$\frac{s_t}{s_l} = \sqrt{\mathrm{Re}_t} = \sqrt{\frac{u' l_t}{\nu}} \tag{3.41}$$

The kinematic viscosity can be assumed to $\nu = D = s_l l_f$ [134], which leads to

$$\frac{s_t}{s_l} = \sqrt{\frac{u' l_t}{s_l l_f}}. \tag{3.42}$$

Following, in small scale turbulence regime, the turbulent flame propagation not only depends on u'/s_l but also on the length scales [135].

Based on theoretical considerations and flame speed measurements, a variety of turbulent flame speed formulations have been developed in the last decades on the basis of Damköhlers findings. A comprehensive survey can be found in [62, 104]. A selection of these formulations will be compared to measured turbulent flame speeds in the following.

Figure 3.25 classifies the turbulent flame speed measurement data of different authors into regimes of turbulent premixed combustion.

The measurement data range from wrinkled flamelets regime to thin reaction zone regime over a wide spread of length scale ratios, i.e. $12 \leq l_t/l_f \leq 290$. Considering the investigations on engine relevant scales in section 3.5, the measurement data almost cover the entire range of premixed turbulent combustion in SI engines.

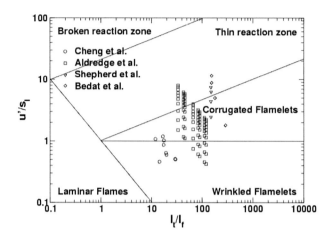

Figure 3.25: Measurement data [8, 17, 30, 160] of turbulent flame velocity in regime diagram of turbulent premixed combustion

3.7.1 Turbulent Flame Speed Formulations based on Velocity Scale

In the following, formulations describing the turbulent flame speed based on the velocity scales are compared with combustion regime classified measurement data in figure 3.26. The formulations are listed in table 3.5 including values for the fitting coefficient C obtained considering the full range of measurement data. As fitting criteria the root mean square error is used.

In the range of wrinkled flamelets the measurement data shows a linear dependency of s_t to u'. The corrugated flamelets regime is characterised by a non-linear slope of s_t/s_l with increasing u'/s_l, called bending effect. In the thin reaction zone regime the bending effect diminishes. For high u'/s_l ratio the dependency can be described as linear.

Formulations A and B describe a linear dependency of s_t/u'. The add-on of the laminar flame speed in formulation B results in an enhancement of s_t/s_l in the wrinkled flamelets regime in comparison to formulation A. In the thin reaction zone regime, the predicted s_t/s_l is in comparison small. Fitted over the full range of measurement data, both formulations fail to predict the measured turbulent flame speeds.

Item	Reference	Formulation	C
A	[139, 140]	$\frac{s_t}{s_l} = C\frac{u'}{s_l}$	3.048
B	[17, 31, 112, 170]	$\frac{s_t}{s_l} = 1 + C\frac{u'}{s_l}$	2.300
C	[85]	$\frac{s_t}{s_l} = C(\frac{u'}{s_l})^{0.7}$	4.159
D	[2, 81]	$\frac{s_t}{s_l} = 1 + C(\frac{u'}{s_l})^{0.7}$	3.759
E	[68]	$\frac{s_t}{s_l} = 1 + C(\frac{u'}{s_l})^{5/6}$	3.055
F	[35]	$\frac{s_t}{s_l} = (0.5(1 + (1 + 8C(\frac{u'}{s_l})^2)^{0.5}))^{0.5}$	448.4

Table 3.5: Turbulent flame speed closure formulations based on velocity scales together with fitting coefficients C obtained considering the fully range of measurement data

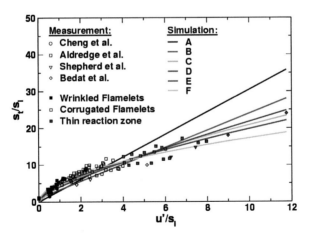

Figure 3.26: Normalised measured [8, 17, 30, 160] and calculated turbulent flame speeds using formulations based on velocity scales listed in table 3.5 as a function of velocity scales

The formulations are comparable to Damköhler's considerations of the corrugated flamelets regime. However, the measured non-linear slope of s_t/s_l with increasing u'/s_l in this intermediate regime can not be predicted using these formulations. Thus, the application range is limited to the wrinkled flamelets regime, in the vicinity of thin reaction zone regime an adequate fitting is needed.

The formulations C and D account for the bending effect in the range of corrugated flamelets with a bending exponent of 0.7. The resulting non-linear slope in this range diminishes with increasing u'/s_l. In the wrinkled flamelets regime, the dependency can be described as linear. Due to the add on of the laminar flame speed in formulation D, the bending effect is much more pronounced in comparison to formulation C, leading to a better match of measurement data in total.

An increased bending exponent up to 5/6 (formulation E[1]) leads to a measurable decrease of bending. The formulation describes rather a linear than a non-linear slope in the range of corrugated flamelets. In formulation F the bending is much more pronounced with an exponent of 0.5. However, bending occurs already in the range of wrinkled flamelets.

Summarising, formulations A and B work well in the range of wrinkled flamelets and thin reaction zone regime, but fail to predict turbulent flame velocities in the corrugated flamelet regime. Formulations C and D predict well the turbulent flame speed in the full range of turbulent premixed SI engine combustion. The add-on of the laminar flame speed in formulation D results in an even better match of measurement data[1]. In comparison to formulations C and D, the bending exponent is in formulations E and F increased. As a result, these formulations fail to predict the turbulent flame speed in the wrinkled and corrugated flamelets regimes. Implying adequate fitting, these formulations can be used for turbulent flame speed calculation in the thin reaction zone regime.

3.7.2   Turbulent Flame Speed Formulations based on Velocity and Length Scale

Turbulent flame speed formulations requiring no fitting and considering flame altering due to velocity as well as length scales are listed in table 3.6. A comparison between normalised formulations with measurement data at constant length scale ratios and fuel-air equivalence ratios is

---

[1]Note that the original formulation of [68] is here simplified in terms of considering fully developed turbulent combustion.

[1]If not indicated otherwise, formulation D is used as closure formulation for the G-equation approach in this thesis.

shown in figure 3.27. Note that the data is normalised using laminar flame speeds of methane-air mixtures measured by Egolfopoulos et al. [47].

Item	Reference	Formulation
A	[140]	$\frac{s_t}{s_l} = \frac{u'}{s_l}(0.25 + \log_{10}(\frac{s_l l_t}{u' l_f}))$
B	[132]	$\frac{s_t}{s_l} = 1 - 0.195\frac{l_t}{l_f} + ((0.195\frac{l_t}{l_f})^2 + 0.78\frac{u' l_t}{s_l l_f})^{0.5}$
C	[132]	$\frac{s_t}{s_l} = (0.78\frac{u' l_t}{s_l l_f})^{0.5}$
D	[155]	$\frac{s_t}{s_l} = 1 + \frac{u'}{s_l}(1 + (\frac{l_t s_l}{u' l_f})^{-2})^{-1/4}$
E	[88, 89]	$\frac{s_t}{s_l} = (\frac{18 C_\mu}{(2C_m-1)\beta'}((2K_c^* - C_\tau C_4)(\frac{u' l_t}{s_l l_{f,th}}) + \frac{2C_3}{3}(\frac{u'}{s_l})^2))^{0.5}$

Table 3.6: Turbulent flame speed closure formulations based on velocity and length scales

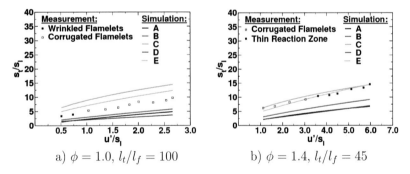

a) $\phi = 1.0$, $l_t/l_f = 100$ b) $\phi = 1.4$, $l_t/l_f = 45$

Figure 3.27: Normalised measured [8] and calculated turbulent flame speeds using formulations based on velocity and length scales listed in table 3.6 as a function of velocity scales for different fuel-air equivalence and length scale ratios

Formulation A, with a linear scaling of u'/s_l and logarithmic impact of the length scale ratio, possesses a nearly linear increase of s_t/s_l with increasing $u^{'}/s_l$. A marginal bending can be observed for high ratios of $u^{'}/s_l$ close to the thin reaction zone regime.

Peters' [132] formulations B and C show a well pronounced bending in the intermediate regime which diminishes in the vicinity of the thin reaction zone regime. The bending is much more pronounced using the formulation describing the turbulent burning velocity in the limiting case of small scale turbulence (formulation C). This formulation matches well for measurement data of fuel-air equivalence ratio 1.4.

Formulation D behaves similar to formulation A in terms of linear increase of s_t/s_l with increasing $u^{'}/s_l$ due to low scaled impact of l_t/l_f and linear impact of $u^{'}/s_l$. Whereat, a bending in the intermediate region can not be observed.

Formulation E by Kolla et al.[1] [88, 89] matches the measurement data for $\phi = 1.4$ well, too. The bending is defined in-line with the measurements for the lower fuel-air equivalence ratio. For increasing ratios of $u^{'}/s_l$ the enhancement of s_t/s_l becomes increasingly linear. Note that in contrast to the other formulations investigated, this formulation is based on the thermal flame thickness $l_{f,th} = (2(1 + C_\tau)^{0.7})l_f$, where the Zeldovich thickness l_f is determined using equation 3.15.

The three formulations B, C and E, offering the most promising approach for modelling the turbulent burning velocity in 2-dimensional, comparison with measurement data will be investigated in more detail in the following. Figure 3.28 displays the functions dependent on length and velocity scales in comparison with measurement data.

In general, all formulations show an increase of s_t/s_l with $u^{'}/s_l$ and l_t/l_f, while the increase of s_t/s_l intensifies with larger l_t/l_f.

Peters' formulation B matches the measurement data in the wrinkled flamelet regime quite well. In the corrugated flamelets regime the turbulent flame speed is slightly underestimated. The underestimation increases with $u^{'}/s_l$ and proceeds in the thin reaction zone regime. Good accordance with measurement data can be found for length scale ratio of 68.2 as well as l_t/l_f values above 150.

Formulation C possesses a good accordance with the measurement data in the range of low l_t/l_f values up to 68.2 for all $u^{'}/s_l$ ratios investigated. With increasing l_t/l_f ratio the formulation increasingly overes-

---

[1]The coefficients here are defined by the authors [88, 89] as following: C_μ as modelling constant from $k - \varepsilon$ model, $C_m = 0.7$, $\beta^{'} = 6.7$ as flamelet curvature contribution term, $K_c^*/C_\tau \simeq 0.85$ as dilatation rate for methane-air mixtures, $C_\tau = (T_b - T_u)/T_u$ as heat release parameter, C_3 and C_4 representing the turbulence-scalar interaction effects dependent on Karlovitz number $\text{Ka} = ((2(1 + C_\tau)^{0.7})^{-1}(u^{'}/s_l)^3(l_{f,th}/l_t))^{0.5}$, with $C_3 = 1.5(\text{Ka})^{0.5}/(1 + (\text{Ka})^{0.5})$ and $C_4 = 1.1(1 + \text{Ka})^{-0.4}$.

timates the measurement data, independently of turbulent combustion regime. The error enhances with increasing u'/s_l.

The formulation of Kolla et al. (formulation E) predicts well the measured turbulent flame speed data in the range of low and medium $l_t/l_{f,th}$ ratios for all u'/s_l values. For $l_t/l_{f,th}$ values greater than 19.5 (which corresponds to a l_t/l_f ratio of 119.6) the formulation overestimates the measurement data.

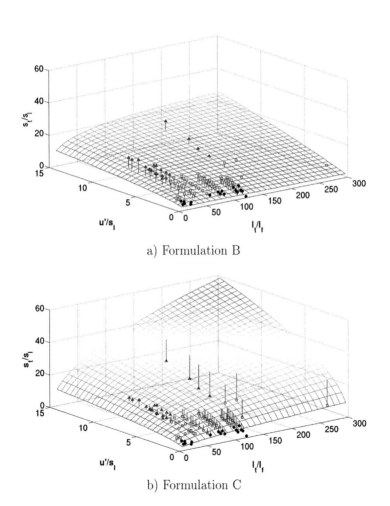

a) Formulation B

b) Formulation C

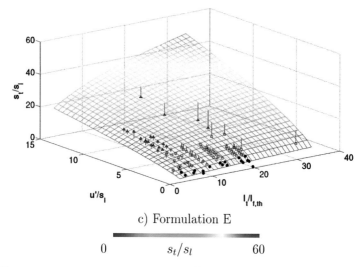

c) Formulation E

0   s_t/s_l 60

Figure 3.28: Normalised measured [8, 17, 30, 160] (symbols) and calculated (contour) turbulent burning velocities as a function of length and velocity scales. Black points refer to wrinkled flamelets regime, white rectangles to corrugated flamelets regime and grey triangles to thin reaction zone regime. Lines between symbols and contour denote difference between measured and calculated s_t/s_l ratio.

Summarising, formulation B works well for low u'/s_l values (< 2) and independently of it for high l_t/l_f values (> 150). Formulation C performs good in the range of small l_t/l_f values (< 68.2) for low and medium u'/s_l. In comparison formulation E matches all measurement data up to l_t/l_f values of 119.6 for most u'/s_l ratios investigated.

3.7.3 Influence of Turbulent Flame Speed Formulation on Pressure Trace in Simplified Test Case

The influence of the turbulent flame speed formulations B, C and E (see 3.7.2) on the pressure curve is investigated using the simplified test case specified in appendix A.1.2 by considering the engine relevant combustion regimes represented in section 3.5. The chamber is homogeneously initialised with engine-relevant conditions at spark timing, i.e. a temperature of 800 K and a pressure of 20 bar. The initial values of u' and l_t

are varied to match the combustion regime characteristics. Figure 3.29 displays the pressure trace calculated using the different turbulent flame speed functions in the wrinkled flamelets, corrugated flamelets and thin reaction zone regime.

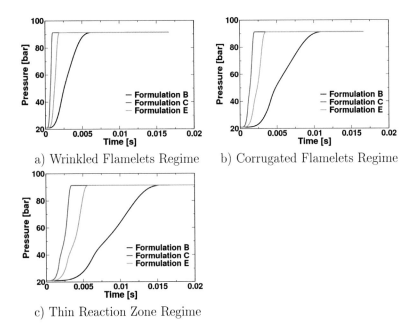

a) Wrinkled Flamelets Regime   b) Corrugated Flamelets Regime

c) Thin Reaction Zone Regime

Figure 3.29: Influence of turbulent flame speed closure formulations on pressure curve calculated in simplified test case for different regimes of turbulent premixed combustion

Independent of the considered turbulent flamelet regime, formulation C shows the greatest gradient of pressure increase while the smallest gradient can be observed for formulation B. In the wrinkled flamelets regime (figure 3.29 a) the different turbulent flame speed formulations reach their maximum chamber pressure in 0.001 s (formulation C), 0.002 s (formulation E) and 0.006 s (formulation B). The spread of reaching the maximum chamber pressure increases with the ratio of u'/s_l. In the corrugated flamelets regime (figure 3.29 b), the spread is in the range of 0.008 s. In the thin reaction zone regime (figure 3.29 c), the difference in reaching the maximum cylinder pressure increases up to 0.0117 s.

These results emphasise the outcome of the 2-dimensional comparison of the different formulations with measurement data displayed in figure 3.28. The different turbulent flame speed functions predict a quite similar flame speed in the wrinkled flamelet regime. In the corrugated flamelets regime, remarkable differences between the calculated flame speeds can be observed. These differences increase strongly in the extend of the thin reaction zone regime.

3.7.4   Influence of Turbulent Flame Speed Formulation on Pressure Trace in Engine Test Case

As shown in section 3.5, flame propagation in premixed turbulent combustion takes place, dependent on engine characteristics and operating strategy, primarily in the thin reaction zone regime as well as corrugated flamelets regime. In both regimes, considerable differences between the calculated turbulent flame speeds can be observed using the formulations B, C and E (see 3.7.2 and 3.7.3).

In order to evaluate the impact of these differences on calculated pressure curve, the different turbulent flame speed functions are investigated on a big mildly turbocharged SI engine at two full load operating points (OP A: 3000 rpm, 17.47 bar and OP B: 5000 rpm, 15.72 bar). For turbulent flame speed closure, the approximation function of Gülder is used. Details about engine specification and operating points can be found in appendix A.4.1 and A.4.2.

Figure 3.30 compares the calculated pressure curves using the different turbulent flame speed functions with measurement data.

Using the turbulent flame speed function B, the pressure increase is in comparison with the measurement data at both operating points retarded. As a result, the calculated maximum in-cylinder pressure underestimates the measured values by 22 bar (OP A) and 11 bar (OP B). Formulation C leads to an overestimation of the measured in-cylinder pressure. At both operating points, the combustion starts considerably before the measured combustion onsets and the maximum in-cylinder pressure differs from the measured values by 78 bar (OP A) and 81 bar (OP B). The calculated in-cylinder pressures using formulation E differ from the measurement data, too. The maximum in-cylinder pressures predicted are 40 bar (OP A) and 37 bar (OP B) higher than the values measured.

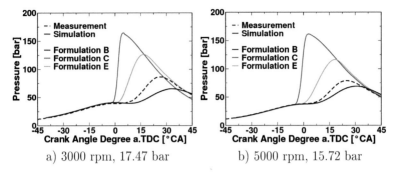

a) 3000 rpm, 17.47 bar   b) 5000 rpm, 15.72 bar

Figure 3.30: Measured and calculated pressure curves using different turbulent flame speed functions at two full load operating points of the big mildly turbocharged engine

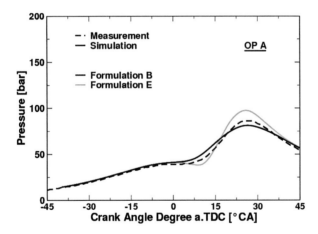

Figure 3.31: Measured and calculated pressure curves using different turbulent flame speed functions at full load operating point (3000 rpm, 17.47 bar) of the big mildly turbocharged engine. In contrast to figure 3.30, the flame kernel diameter is adjusted for both formulations in order to match the in-cylinder pressure trace.

However, the G-equation model is sensitive towards the flame kernel modelling approach. By tuning the engine-specific kernel diameter, the combustion onset can be adjusted[1]. Figure 3.31 displays the calculated pressure curves obtained by adjusting the kernel diameter for formulation B and E[2].

The figure reveals, that the overall shape of the calculated pressure curve changes by adjusting the kernel diameter only marginally. As a result, formulation B slightly underestimates the measured pressure curve, and formulation E slightly overestimates the in-cylinder pressure.

Thus, for predictive premixed SI engine turbulent flame propagation modelling, formulations B and C constitute promising approaches. However, for a better reproduction of the measured pressure curves a more accurate laminar flame speed fitting function is required.

3.8 Model Validation using Optical Measurements

In the following, the flame front propagation calculated using the G-equation model is validated with optical measurements on the highly turbocharged SI engine. In order to track the mean flame front, the measurement system utilises OH radical chemiluminscence. The endoscope used for the measurements accesses the combustion chamber through the cylinder head. The visual range inside the combustion chamber is depicted in figure 3.32. Details about the measurement system can be found in appendix A.2.3.

The G-equation model is used with the turbulent flame speed formulation D introduced in section 3.7.1 and the laminar flame speed fitting function of Perlman. The turbulent flame speed closure demands for a fitting of the coefficient C, which is done for the individual cases based on the burning rates measured[3].

In the following, a variation of the global fuel-air equivalence ratio ($\phi = 0.9, 1.0, 1.3$) at part load operating point at 2000 rpm and 5 bar of the highly turbocharged SI engine is investigated. In test bed measurements, the 50 % transformation point is kept constant by adjusting the spark timing. Details about the operating point can be found in appendix A.2.2. The calculated in-cylinder pressure curves are compared with the measurement data in figure 3.33.

---

[1]Typically the kernel diameter is varied between 0.001 m and 0.004 m.

[2]Due to large differences between calculated and measured in-cylinder pressure traces which become apparent from figure 3.30, formulation C is excluded from investigation below.

[3]For detailed information see [106].

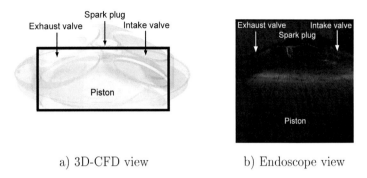

a) 3D-CFD view   b) Endoscope view

Figure 3.32: Visual range inside the combustion chamber of the highly turbocharged SI engine

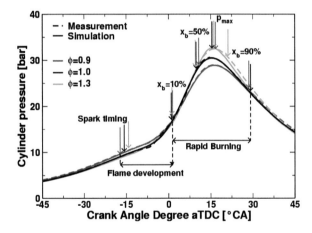

Figure 3.33: Measured and calculated pressure curves of highly turbocharged engine at part load operating point (2000 rpm, 5 bar) for varying fuel-air equivalence ratios

Additionally, the figure reveals the characteristic combustion timings, i.e. spark timing, 10 %, 50 %, and 90 % transformation point as well as time of maximal in-cylinder pressure. Following Heywood [70], these timings can be used to classify the combustion process into two phases: the flame development phase starting at spark timing and the rapid burning phase, following up the 10 % transformation point.

The flame development phase of the fuel-air equivalence ratio variation is characterised by different spark timings in the total range of 3 crank angle degrees, following the ranking $t_{Spark,\phi=0.9} < t_{Spark,\phi=1.0} < t_{Spark,\phi=1.3}$. However, the time of the 10 % transformation point is comparable for all fuel-air equivalence ratios investigated. In the rapid burning phase differences occur at the point of 90 % transformation. Here, the global $\phi = 1.3$ possesses its characteristic time earlier than $\phi = 0.9$ and $\phi = 1.0$. Additionally, the curves differ in terms of maximal in-cylinder pressure, following the ranking of spark timing.

In general, the calculated pressure curves match the measurement data well. Small differences can be observed in the rapid burning phase after the point of maximal in-cylinder pressure for $\phi = 1.3$.

The calculated flame propagation is compared with chemiluminescence measurements in the following. Note, that the measured flame occurs to be close to the limit of visible range already in an early stage of combustion ($t < t_{x_b=10\%}$). This is due to the wide-angle effect of the endoscope, which can be seen in figure 3.32 and due to flame propagation in direction of the endoscope. For this reason, only the flame development phase is considered in the following. Figure 3.34 depicts the predicted and measured flame propagation in the flame development phase of the fuel-air equivalence ratio variation.

At spark timing the chemiluminescence measurements show the arc-over of the spark plug. The arc-over is still present at 4 °CA bTDC in case of $\phi = 1.0$ and 1.3. Since no detailed flame kernel development model is used in the 3D-CFD simulation, the ignition process of the mixture at the spark plug can not be predicted.

At 6 °CA bTDC a marginal illumination can be seen in the measurements close to the spark plug, which can be related to the arising flame. Here, the cases $\phi = 1.0$ and $\phi = 1.3$ illuminate in a higher degree than $\phi = 0.9$, which can be related to the flame size. A flame shape can not be identified at this stage from the measurement data. The corresponding calculations of the propagating flame show at this stage of combustion an initial flame. The size of the calculated flame can be ranked according to $\phi = 0.9 < \phi = 1.0 < \phi = 1.3$. While the calculated flame of $\phi = 1.3$ occurs to be slightly kidney-shaped in direction of the exhaust valves, the $\phi = 0.9$ and 1.0 flames can be described as spherical.

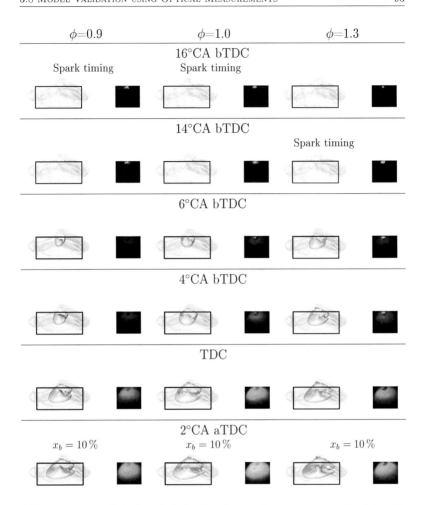

Figure 3.34: Calculated (left, blue iso-surface represents G=0 level) and measured (right) flame propagation in the flame development phase of highly turbocharged engine at part load operating point (2000 rpm, 5 bar) for varying fuel-air equivalence ratios

The measurement data shows an increased flame volume at 4 °CA bTDC. Still, the size of the measured $\phi = 0.9$ flame is smaller compared to $\phi = 1.0$ and 1.3. At this stage, the chemiluminescence is intensified, thus enabling an observation of flame shape. The measured $\phi = 0.9$ and 1.0 flames occur to be spherical. Whereas, the $\phi = 1.3$ flame possesses a kidney-shape in direction of the exhaust valves. The volume of the calculated flames is increased, too. In line with the measured flame shapes, the calculated $\phi = 0.9$ and 1.0 flames occur to be spherical while the $\phi = 1.3$ flame is kidney-shaped.

At TDC, the measured flames show a further volume increase. The $\phi = 1.0$ and 1.3 flames are already close to the limit of visible range making an analyse in terms of total flame volume difficult. Whereas, the $\phi = 0.9$ flame still possesses a remarkable distance from the limit of visible range. Thus, its size is smaller compared with $\phi = 1.0$ and 1.3. From measurements it can be further observed, that the $\phi = 1.3$ flame intensified its kidney-shape. The $\phi = 0.9$ and 1.0 flames show a kidney-shape at this stage, too. Compared with the $\phi = 1.3$ flame their shape occurs to be less edged. The calculations show an explicit ranking of the flame sizes, following $\phi = 0.9 < \phi = 1.0 < \phi = 1.3$. In line with the measurements, the $\phi = 0.9$ and 1.0 flames occur to be kidney-shaped but less pronounced than the $\phi = 1.3$ flame.

At 2 °CA aTDC the flames measured have reached the limit of visual range. Their chemiluminescence intensity differs, which can be attributed to the different pronounced flame propagation in direction to the endoscope. From the calculations it can be concluded, that at 10 % transformation point the flames show the same shape in general. However, their wrinkling degree differs strongly, following the ranking $\phi = 0.9 < \phi = 1.0 < \phi = 1.3$ and thus their related homogeneity, already described in section 3.6.5.

Summarising, the calculated flames of the fuel-air equivalence ratio variation match the optical measurements of the flames in terms of shape and local propagation speed well. Thus, the G-equation model is in combination with a simplified turbulent flame speed formalisms and the Perlman laminar flame speed closure formulation a tool, enabling the modelling of turbulent premixed flame propagation incorporating turbulent and chemical effects.

Chapter 4

Auto-Ignition Modelling

4.1 Introduction

Auto-ignition processes are strongly dependent on fuel type and thermo-dynamic state. The auto-ignition occurs after a certain ignition delay time, in which chemical pre-reactions form radicals, initiating the fuel transformation process. One attempt to quantify the fuel-specific ignition delay time is to determine its Research Octane Number (RON) or Cetane Number (CN) under defined conditions, whereat a high RON and a low CN characterise fuels possessing a long ignition delay time[1]. In general, the fuel specific ignition delay time is decreased with increasing temperature and pressure.

The usage of low RON or rather high CN fuel enables the operation of conventional diesel engines. The combustion process bases thereby on the fast auto-ignition of the fuel mixed with air by convection and diffusion. Whereas, conventional SI engines operate with high RON fuels. The transformation of the in-cylinder mixture is here initiated by a spark, igniting a premixed flame kernel, which grows to a premixed flame, propagating through the combustion chamber.

The pressure gradients occurring due to auto-ignition are greater than the ones appearing in premixed flame combustion. In compari-son with SI engines, diesel engines are therefore constructed with higher thermal and mechanical strength, allowing engine operation based on auto-ignition. However, auto-ignition phenomena can also occur in SI engines. Due to the lower thermal and mechanical strength, the phenom-ena are strictly undesired, since they can cause serious engine damage.

The auto-ignition phenomena occurring in SI engines can be clas-sified in engine knock, pre-ignition (also referred as super knock), and

---

[1]The RON and CN are determined according to DIN EN 228 and DIN EN 5165, re-spectively.

homogeneous ignition. Engine knock is a result of unburnt mixture compression due to flame propagation. The compression of the unburnt mixture leads to a local increase of temperature and pressure. In case the resulting decreased ignition delay time is shorter than the time the flame requires to reach and transform the mixture locally, auto-ignition occurs. Since this phenomena is directly linked to flame propagation, it can be controlled by spark timing shift. Pre-ignition phenomena occur stochastically before spark timing under heavily enhanced boost pressures at low engine speeds. Since the auto-ignition process of the unburnt mixture starts out before spark timing, a controlling by spark timing shift is not feasible. Some SI engines base on the auto-ignition of the premixed mixture, like HCCI engines. In order to initiate the auto-ignition process, a high temperature level is generated in the end of compression stroke. An effective way to achieve a temperature enhancement is to use the exhaust gases from the previous combustion cycle. Using these, the specific heat capacity of the in-cylinder charge is enhanced, and the pressure gradient resulting from the homogeneous auto-ignition process is reduced.

In order to model the variety of SI engine auto-ignition phenomena in 3D-CFD simulations, the fuel-specific chemistry needs to be incorporated. The conflict of goals arising in coupling 3D-CFD calculations with detailed chemistry is to keep computational costs low while achieving accurate results. A compromise in this trade-off constitutes the tabulation of the chemistry, as done in the Ignition Progress Variable approach. However, the results obtained applying tabulation methods depend strongly on the accuracy of tabulation.

Following a brief literature survey on auto-ignition modelling, the IPV model is introduced in section 4.3. The coupling of the IPV model with the 3D-CFD code is described in section 4.4. The accuracy of IPV library is investigated in section 4.5. To model the variety of auto-ignition phenomena, the IPV model needs to be coupled with the flame propagation model. The functionality of the coupling is investigated by means of an HCCI engine in section 4.6. Afterwards, the capability of this approach to predict pre-ignition phenomena is examined for a highly turbocharged SI engine in section 4.7.

4.2   Literature Survey

In the literature different approaches for auto-ignition modelling are described. Basically these approaches can be classified in approaches incorporating detailed chemistry and approaches which do not account for detailed chemistry[1]. The most important models are listed in table 4.1 along with the problem addressed in model application.

Simplified auto-ignition models incorporate fuel-specific chemistry in 3D-CFD simulations based on a limited number of reaction steps. Examples for these models are the Livengood-Wu integral [101,105] and the Shell auto-ignition model and its derivatives [20,24,46,64]. The last one describes the fuel specific chemistry already in marginal detail.

Approaches incorporating detailed chemistry base on well validated detailed chemical reaction mechanism, consisting of several hundreds of chemical species and thousands of reactions. The integration of the reaction mechanism in 3D-CFD can be done based on different approaches, i.e. directly coupled, tabulated, or decoupled.

The direct integration of the detailed chemistry implies the transport of each chemical species of the detailed reaction mechanism and the in-situ solution of chemistry on cell level. Considering the number of species a reaction mechanism is build of, the direct integration implies a very high computational demand. For this reason, authors adopting this approach only apply a strongly reduced reaction mechanism [28,102,162,192]. Some authors [92] limit their investigations on sector meshes only. However, the direct integration is the most straight forward and accurate way of coupling detailed chemistry with 3D-CFD simulations.

Another approach for integrating detailed chemistry in 3D-CFD simulations is the tabulation of the detailed chemistry solution in form of a representative variable. The solution of the detailed chemistry is either tabulated in-situ for a given set of independent variables, as done in the In-Situ Adaptive-Tabulation (ISAT) model [140], or in advance for a wide range of independent variables. The last method is applied in the Tabulated Kinetics of Ignition (TKI) [87] model, the reaction trajectory model of Mass et al. [152], and the Flamelet Progress Variable (FPV) model [99]. The tabulation approach only demands for the solution of transport equations for the representative variables, thus keeping the computational demand low, while allowing the consideration of local effects, like variations of fuel-air equivalence ratio and turbulence, on auto-ignition chemistry on cell level.

---

[1]A further classification can be made in terms of accounting for turbulence and chemistry interaction. Assuming an almost homogeneous distribution of the mixture in the area of auto-ignition, the influence of the flow on the chemistry will not be considered in the following.

Method	Tool	Reference	Problem Description
Reduced Chemistry	Livengood-Wu Integral	[105]	HCCI Combustion [101]
	Shell	[64]	Engine Knock [20, 24, 46]
Direct Integration of Detailed Chemistry			Engine Knock [28, 78, 156], HCCI Combustion [92, 162]
In-Situ Tabulation of Detailed Chemistry	ISAT	[141]	HCCI Combustion [37, 48]
Pre-Processing Tabulation of Detailed Chemistry	TKI	[86]	HCCI Combustion [87]
	Maas-Model	[152]	HCCI Combustion [152]
	FPV	[99]	Diesel Engine Combustion [99, 100]
Indirect Integration of Detailed Chemistry	Multi-Zone	[3]	HCCI Combustion [3, 5, 76, 157, 174] [4, 98, 153, 175]
	Kinetic Maps		Engine Knock [15], Pre-Ignition [188], HCCI Combustion [187]

Table 4.1: Tools for auto-ignition modelling and application

Approaches basing on the indirect integration of the detailed chemistry solve the flow and mixture formation problem in 3D-CFD decoupled from detailed chemistry. In the multi-zone model [3], the 3D-CFD flow and mixture formation solution is divided in several groups (so-called chemical clusters) based on certain criteria, like temperature or fuel-air equivalence ratio. The characteristic conditions of each group are averaged and provided as input values to a chemistry solver. This method works effective when large areas of the combustion chamber are homogeneous in terms of composition and enthalpy. Another approach basing upon a decoupled solution of the 3D-CFD and chemistry solution is the kinetic mapping. Herein, variables describing the mixture characteristic of the 3D-CFD solution are mapped on kinetic maps resulting from chemistry solution in post-processing. This approach offers a phenomenological prediction of engine knock, pre-ignition and HCCI combustion [15, 187, 188], and constitutes the cheapest way in terms of computational costs to model auto-ignition phenomena in SI engines.

Summarising, the direct integration of detailed chemistry in the 3D-CFD code is the most accurate way but involves high computational costs. The indirect method keeps the computational costs low but allows only a phenomenological prediction of auto-ignition processes. A compromise in the trade-off achieving accurate results by keeping computational costs low is the tabulation of the detailed chemistry solution in form of a representative variable.

4.3   IPV Model Description

From diesel engine applications [99, 100, 117], it is known that the auto-ignition process from not ignited, unburned state, to steady, auto-ignited state can be described based on a tabulated Ignition Progress Variable (IPV) (also referred as FPV), which is transported in 3D-CFD code in order to facilitate interaction with the library.

Lehtiniemi [99, 100] first suggested to base the reaction progress variable on the latent enthalpy, i.e. the enthalpy at standard state $h_{T=298K}$, which is integrated over the flamelet [99, 100, 117]. As the flamelet ignites, the species composition changes at different locations in the mixture fraction space, and the integrated latent enthalpy builds a monotonic function of time. Thus, the integrated enthalpy is a bijective relation between the ignition progress and the flamelet time [99]. Assuming a homogeneous mixture, the integration over mixture composition can be neglected and the ignition progress variable reads

$$c(t) = \frac{h_{T=298K}(t) - h_{T=298K}(t_0)}{h_{T=298K}(t_\infty) - h_{T=298K}(t_0)} \qquad (4.1)$$

with

$$h_{T=298K} = \sum_{i=1}^{N} Y_i h_{i(T=298K)} \qquad (4.2)$$

where $h_{T=298K}(t)$ corresponds to the actual latent enthalpy, $h_{T=298K}(t_0)$ to the enthalpy at initial state, and $h_{T=298K}(t_\infty)$ to the enthalpy at equilibrium state. Y_i defines the mass fraction of species i.

The progress variable describes the chemical evolution of an igniting flamelet [99], thus providing the source term to be stored in the library as a function of independent variables combustion progress c, pressure p, enthalpy h and EGR mass fraction ψ.

$$\dot{c}_{Chemistry}(c, \phi, p, h, \psi) = \dot{\omega}_{Chemistry} \qquad (4.3)$$

The progress variable transport equation can be achieved by a co-ordinate transformation to mixture-fraction progress variable space of the enthalpy equation and requires only the source term of c itself to be stored in the library [99]. A detailed derivation of the transport equation can be found in [100].

In general, the representative progress variable can base on state variables as well as specific combustion species. The ability to describe the reaction progress requires a direct interrelation between the specific variable and overall auto-ignition progress as well as a continuity of the variable, or alternatively combustion species.

The description of the chemical evolution by the integrated latent enthalpy is demonstrated in the following by considering the hydrocarbon fuel specific low (LT) and high temperature (HT) reaction pathways, illustrated in figure 4.1. Both pathways are outlined in the following. Detailed descriptions can be found in [7, 39, 178, 184, 185]. Afterwards, alternative progress variable approaches are compared with the latent enthalpy approach.

The fuel oxidation process is initiated by the low temperature oxidation pathway for temperatures smaller than 700 K. Following an initialisation reaction $RH + O_2 \rightarrow R + HO_2$, the alkyl radicals[1] R are transformed to alkyl peroxy radicals RO_2. Subsequently, the alkyl per-

---

[1]Following Curran et al. [39], alkyl groups are abbreviated with symbol R, alkanes are declared with RH and Q denotes C_nH_{2n} species or structures.

oxy radicals form alkoxyl radicals RO, which dissociate to more stable and smaller alkyl radicals, olefine and combustion products. The main reaction path constitute isomerisation reactions ($RO_2 = QOOH$). The QOOH species are oxygenated to O_2QOOH, forming ketohydroperoxides OQOO through an internal abstraction of H via OQOOH. Latter one promotes chain branching reactions and thus accelerates the global reaction progress. The ketohydroperoxides dissociate to combustion products and OH via CH_2O and HCO.

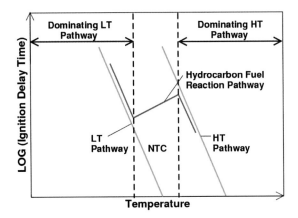

Figure 4.1: Schematic illustration of hydrocarbon fuel specific reaction pathways according to [158]

The lower bound of the high temperature pathway is, dependent on the fuel, around $T = 850, ..., 1000$ K [7]. The oxidation process is initiated by unimolecular fuel decomposition reactions $RH \rightarrow R + X$ (with $X = H, CH_3$) forming alkyl radicals R. Additionally, the fuel is decomposed to R by radicals (OH, HO_2) formed in chain branching reactions. Subsequently, the alkyl radicals dissociate to smaller species and olefine. The olefine formed are decomposed by radicals in smaller olefine and combustion products.

Low as well as high temperature chemistry are characterised by a decreasing ignition delay time with increasing temperature. However, the region in between the low and high temperature kinetic pathways is characterised by an inverse dependency of the ignition delay time to the temperature, and is called region of Negative Temperature Coefficient

(NTC). The low temperature chemistry species O_2QOOH dissociates at high temperatures due to its instability forming the reactants $QOOH$ and O_2. In a chain termination reaction $QOOH$ dissociates forming HO_2 and H_2O_2, which dissociates further forming OH radicals. As a result, a temperature enhancement suppresses the abstraction reactions and subsequent chain branching reactions.

In the following, the development of the integrated latent enthalpy is compared with the course of species characterising the fuel oxidation process in 0D homogeneous reactor calculation. As representative species the following are chosen: $RH = I\text{-}C_8H_{18}$, $R = L\text{-}C_8H_{17}$, $RO_2 = L\text{-}C_8H_{17}O_2$, $RO = L\text{-}C_8H_{17}O$, $QOOH = A\text{-}OOH\text{-}I\text{-}OCT\text{-}D$, $O_2QOOH = A\text{-}OOH\text{-}I\text{-}OCTO2\text{-}D$, $OQOO = A\text{-}O\text{-}I\text{-}OCTOOH\text{-}D$, olefin $= C_4H_8$, and aldehyde $= CH_2O$. Figure 4.2 displays the species profiles and the development of the integrated latent enthalpy.

At an initial temperature of 600 K, the dc/dt profile possesses three maxima. The first maximum is characterised by an initial increase of all intermediate species and combustion products, except for olefin. In this stage of combustion, the major part of fuel RH is transformed while only a marginal amount of CO and CO_2 are formed. Close to the first maximum the mass fractions of the intermediates R, RO, RO_2, its isomer species $QOOH$, O_2QOOH, $OQOO$, and consequently aldehyde start to decrease. Thus, this first ignition stage is a result of LT chemical reactions. The absence of olefin production indicates that this stage of combustion is dominated rather by isomerisation reactions than by a decomposition of RO_2 via RO.

Preceding to the second ignition stage, the dc/dt profile reveals a local minimum. The minimum is characterised by a change of strong decrease to marginal decrease of R and a $QOOH$ species mass fraction close to zero. At the same time HO_2 and H_2O_2 possess a local maximum. Hence, the local minimum of the dc/dt profile is a result of the chain termination reactions $QOOH \rightarrow HO_2 \rightarrow H_2O_2$ in the region of NTC.

Until reaching the second maximum, R decreases only marginally while at the maximum its mass fraction becomes zero. At the same time the HO_2 and aldehyde mass fractions increase continuously reaching their maxima simultaneously with dc/dt. Meanwhile, the mass fractions of H_2O_2, olefine, CO and CO_2 increase. Thus, the second maximum of dc/dt can be attributed to the transformation of R forming aldehyde and HO_2 radicals.

The HO_2 radicals formed are initially dissociated to OH radicals accelerating the combustion process in the third stage of auto-ignition. The aldehyde formed in the second stage of ignition is transformed and a large amount of olefin is produced. Due to the increased temperature, large amounts of radicals are formed in chain branching reactions in the following and thus the HO_2 mass fraction increases again. The high concentration of radicals accelerates the transformation of olefin into CO and consecutively CO_2 in this HT pathway ignition stage.

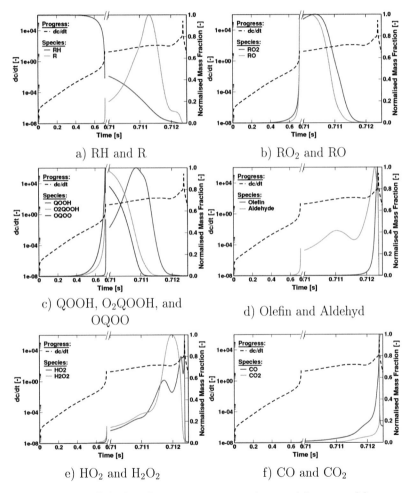

Figure 4.2: Calculated species concentrations and integrated latent enthalpy as a function of time for $\phi = 1.0$, $T = 600$ K, $p = 20$ bar and $\psi = 0.0$

The maxima of dc/dt and the associated species transformations described can be related to phenomenological combustion stages known from literature, i.e. the cool flame, the blue flame, and the main heat release. With increasing temperature, the high temperature oxidation pathway dominates continuously. The increase of dc/dt due to low temperature chemistry is thus less pronounced until it disappears entirely and the dc/dt profile shows only one maximum, the main heat release.

Accounting for progress variable requirements of direct interrelation between progress variable and overall auto-ignition progress as well as continuity, figure 4.2 indicates that a progress variable based on an intermediate species is not adequate enough. However, the continuity requirement is fulfilled by the products of combustion, i.e. CO and CO_2. Some progress variable models known in the literature [152, 167] base on these species. Another state variable fulfilling both requirements is the entropy. In figure 4.3 the alternative progress variables are compared with the enthalpy based progress variable.

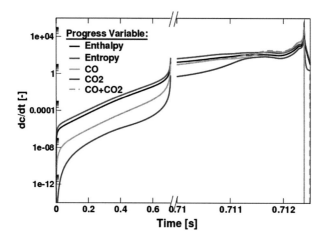

Figure 4.3: Comparison between integrated progress variables based on state variables as well as combustion species

The progress variable based on the integration of CO mass fraction represents the three maxima of dc/dt shown by the enthalpy based progress variable, too. However, the first maximum is shifted in time. Until reaching the first maximum, the variable possesses much smaller

dc/dt values than the enthalpy based progress variable. Thereafter, the dc/dt values of the CO based progress variable outreach the values of the enthalpy based progress variable slightly.

In terms of dc/dt maxima characteristic, the progress variable based on the species CO_2 matches the maxima represented by the enthalpy based progress variable well. In analogy to the CO based variable, the dc/dt values of this variable are much smaller than the one of the enthalpy based progress variable in the period of first ignition stage. After the first ignition stage, the dc/dt values of the CO_2 based progress variable are in the same order of magnitude as the one of the enthalpy based variable.

The integration of the sum of CO and CO_2 mass fraction results in the first and third ignition stage in a progress variable profile similar to the one of the CO integration. Differences of the sum of CO and CO_2 based progress variable to the individual CO and CO_2 based progress variables occur in the second ignition stage. Here, the ignition stage of the integrated sum of CO and CO_2 is extended.

The integrated progress variable based on the entropy possesses a similar characteristic as the progress variable based on the latent enthalpy. The position of the dc/dt maxima as well as the absolute values (apart from a marginal offset) agree well with the one of the enthalpy based progress variable. However, the usage of entropy based progress variable approach demands for introduction of entropy based balance equations in 3D-CFD.

Summarising, the enthalpy as well as the entropy based progress variables represent the overall auto-ignition progress well. The combustion species based progress variables CO and CO_2 are adequate enough to predict the second and third ignition stage. However, the first ignition stage is described by too low dc/dt values especially in the long lasting initial phase[1], which can cause numerical problems. Besides, the timing of the first ignition is represented by the CO based progress variable inaccurately. Thus, to account for all stages of auto-ignition process, the usage of state variables is mandatory.

---

[1]To represent the first ignition stage, CO and CO_2 based progress variable approaches demand for additional consideration of a characteristic first ignition stage combustion species, like CH_2O.

4.4 Coupling to 3D-CFD Code

In addition to the usual 3D-CFD transport equations introduced in chapter 2 the IPV model requires the transport of the progress variable c [21].

$$\bar{\rho}\frac{\partial \tilde{c}}{\partial t} + \bar{\rho}v\frac{\partial \tilde{c}}{\partial x} - \frac{\partial}{\partial x}\left(\bar{\rho}D_t\frac{\partial \bar{c}}{\partial x}\right) = \overline{\rho\dot{\omega}}_{Chemistry} + \overline{\rho\dot{\omega}}_{Spray} \qquad (4.4)$$

In equation 4.4 v describes the velocity, x the spatial coordinate, ρ the density, D_t the turbulent diffusion coefficient and $\dot{\omega}$ the source terms of chemistry and spray evaporation. The chemical source term is obtained from the IPV library according to equation 4.3. The dissipation of c is neglected, thus its variance equation does not need to be solved.

Figure 4.4 illustrates the coupling scheme between the IPV library and the 3D-CFD code.

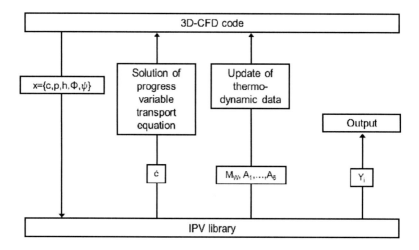

Figure 4.4: Coupling scheme between IPV library and 3D-CFD solver

The 3D-CFD solver calls the IPV library at each time step for all cells in the CFD grid. In order to identify the state of homogeneous auto-ignition progress, the local values of pressure p, total enthalpy h, fuel-air equivalence ratio ϕ, EGR rate ψ, and combustion progress c are required. After identifying the state of the homogeneous auto-ignition progress, the source term of c is returned to the 3D-CFD solver.

The updated progress variable $c_{t+\Delta t}$ reads

$$c_{t+\Delta t} = c_t + \dot{c}_t \Delta t \qquad (4.5)$$

$$\overline{c}_1 = \overline{c}_0 + f_{\overline{c}} \left(\dot{\overline{c}} + \frac{\partial \overline{c}}{\partial \psi} \Delta \psi + \frac{\partial \overline{c}}{\partial \phi} \Delta \phi + \frac{\partial \overline{c}}{\partial p} \Delta p + \frac{\partial \overline{c}}{\partial T} \Delta T + \frac{\partial \overline{c}}{\partial c} \Delta \overline{c} \right) \Delta t \qquad (4.6)$$

where $\overline{c}_0$ and $\overline{c}_1$ are the combustion progress variables at the old and new time step. The introduction of the weighting coefficient $f_{\overline{c}}$ allows for the minimisation of tabulated c nodes (see 4.5.2). The weighting coefficient reads for $0 < f_{\overline{c}}^0 < 1$:

$$f_{\overline{c}} = \left(1.875 \overline{c^2} - 2.875 \overline{c} - 1.0 \right) f_{\overline{c}}^0 + \left(1.25 \overline{c} - 0.25 \right) \overline{c}_0. \qquad (4.7)$$

For $f_{\overline{c}}^0 = 0$ the term $f_{\overline{c}}$ is set to zero and for $f_{\overline{c}}^0 = 1$ the term $f_{\overline{c}}$ equals one. The default value of f_c reads 1. The manipulation in the range $0 < f_{\overline{c}}^0 < 1$ changes the integral of the progress variable primarily in the range of $0 < c < 0.5$, as illustrated in figure 4.5.

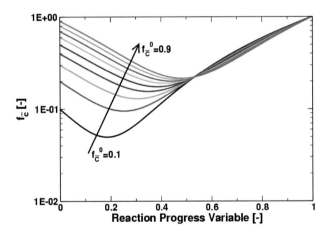

Figure 4.5: Dependency of $f_{\overline{c}}$ on $f_{\overline{c}}^0$ as a function of progress variable c

The weighting coefficient has a parabolic nature. The minimum of the function is shifted towards higher reaction progress variables with increasing $f_{\overline{c}}^0$, while the absolute values of the minimum point are increasing.

From the solved c-field, updated thermodynamic data from the IPV library are used to determine the new temperature and pressure values. In the IPV library a term of the total enthalpy (thermal and chemical enthalpy) is stored as a function of fuel-air equivalence ratio ϕ, temperature T and EGR rate ψ. The NASA-polynomial coefficients [54] of the mixture can be determined from the coefficients of the species $a_{j,i}$ and the mass fractions Y_i

$$A_j = \sum_i Y_i \frac{\overline{M}_W}{M_{W,i}} a_{j,i} \qquad 1 \leq j \leq 6 \qquad (4.8)$$

which are stored in the IPV library among the mixture mean molecular weight $\overline{M}_W$.

$$\overline{M}_W = \left(\sum_i \frac{Y_i}{M_{W,i}} \right)^{-1} \qquad (4.9)$$

The temperature can be determined from the total enthalpy in an iterative procedure.

$$\frac{h}{R} = A_1 T + \frac{A_2}{2} T^2 + \frac{A_3}{3} T^3 + \frac{A_4}{4} T^4 + \frac{A_5}{5} T^5 + A_6 \qquad (4.10)$$

$$\frac{c_p}{R} = A_1 + A_2 T + A_3 T^2 + A_4 T^3 + A_5 T^4 \qquad (4.11)$$

The pressure can be calculated using the ideal gas law.

$$\overline{p} = \frac{\overline{\rho} R T}{M_W} \qquad (4.12)$$

$$R = \frac{R_{Universal}}{M_W} \qquad (4.13)$$

Alternatively, the tabulated temperature and specific heat capacity can be used. Additionally, defined species mass fractions Y_i can be stored in the IPV library, allowing the species output decoupled from the 3D-CFD solution.

In order to model auto-ignition as well as spark plug initiated combustion, the G-equation as well as IPV model are used. Both models modify the overall reaction progress in the combustion chamber. In order to avoid an unrealistic flame front acceleration due the diffusivity of the combustion progress variable, the combustion progress variable is treated differentially dependent on G. For a cell with $G \leq -0.5 l_{f,t}$, i.e. not reached by the flame front, the combustion progress variable transport is calculated by the standard convection flow solver. For a cell with $G > -0.5l_{f,t}$, i.e. reached by the flame front, the combustion progress variable transport is deactivated.

4.5 IPV Library Definition

The IPV library provides significant variables describing the auto-ignition process, tabulated as a function of the combustion progress variable c. The variables describing this process are the time derivative of the combustion progress variable dc/dt and the thermodynamic data tabulated in terms of mean molecular weight M_W and NASA-polynomial coefficients A_i or alternatively specific heat capacity c_p and temperature T. In order to create an IPV library following steps are required:

- Definition of library extension

- Definition of independent variable nodes

- 0D homogeneous reactor calculation,

- Tabulation of data

The definition of the library extension and the independent variable nodes determine the library extension and thus the computational demand for library creation. Both are investigated in the following.

4.5.1 Definition of Library Extension

In the beginning of IPV library creation the library size needs to be defined. The tabulated data must cover the whole state of independent variables describing the thermodynamic state and mixture composition of the engine cycle considered.

The library size can be minimised by examining the reaction progress in dependence on the independent variables. Figures 4.6, 4.7, 4.8, and 4.9 display the dc/dt progress as a function of c for the independent variables fuel-air equivalence ratio ϕ, temperature T, pressure p, and EGR rate ψ.

In figure 4.6 at low c values ($c < 0.1$) a first peak of dc/dt can be observed, which increases and moves towards higher c values with increasing ϕ. The peak enhances due to increasing pre-reactions and thus heat-release with fuel enrichment in this so-called cool flame stage of ignition. The shift in direction of higher c values is a result of decreasing reactivity of the mixture due to increasing specific heat capacity. The second peak refers to the blue flame ignition stage which is directly followed up by the main heat release. With increasing fuel enrichment as well as fuel enleanment the impact of the second stage of ignition on mixture transformation increases strongly. While under stoichiometric conditions approximately 25 % of the mixture are transformed in this stage, the transformation increases up to 70 % under fuel rich conditions ($\phi = 3.4$). For $\phi = 3.8$ the dc/dt profile is in all ignition stages almost comparable to the one of $\phi = 3.4$. Thus, for library minimisation a limitation of fuel-air equivalence ratio to $\phi \leq 3.4$ is valid accepting a marginal error. The limitation assumes the allocation of a variable ϕ_i according to

$$
\phi_i = \begin{cases} \phi_{Tab,min} & \text{for } \phi_i \leq \phi_{Tab,min} \\[2ex] \phi_{Tab,max} & \text{for } \phi_i \geq \phi_{Tab,max} \\[2ex] \phi_i & \text{else.} \end{cases} \tag{4.14}
$$

On the fuel-lean side, the dc/dt profiles of the different fuel-air equivalence ratios are not comparable to each other. Thus, a library minimisation is not feasible on the fuel lean side.

The dc/dt profiles of varying temperatures, as shown in figure 4.7, possess the low and high temperature chemistry influence on reaction progress. Primarily the curves differ in terms of ignition stages apparent (one stage for $T = 1200$ K, two stages for $900 \leq T \leq 1100$ K, and three stages for $500 \leq T \leq 800$ K) and their location in time. The slope of decrease of the dc/dt profile in the range of $0.8 < c < 0.9$ is comparable for all temperatures. However, the absolute values differ, making a library limitation not feasible. Note, that the temperatures $T = 300$ K and $T = 400$ K do not show a reaction progress in the time considered. Thus, a library limitation is valid to $T \geq 400$ K.

The pressure variation depicted in figure 4.8 possesses a shift in start and duration of the ignition stages towards higher c and time values with increasing pressure. The shift is less pronounced with increasing pressure. Hence, for high pressure values ($p = 85$ bar and $p = 95$ bar), the dc/dt profiles of the second and third ignition stage are almost comparable. However, comparable values of dc/dt are not achieved in the first stage of ignition.

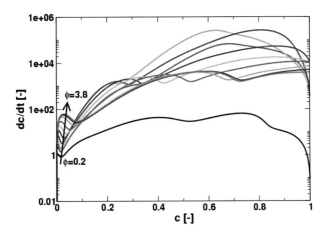

Figure 4.6: Calculated dc/dt profiles as a function of c for $0.2 \leq \phi \leq 3.8$ ($\Delta\phi = 0.4$), $T = 800$ K, $p = 20$ bar, and $\psi = 0.0$

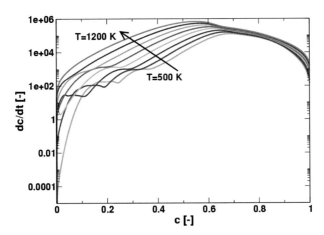

Figure 4.7: Calculated dc/dt profiles as a function of c for $\phi = 1.0$, $500 \leq T \leq 1200$ K ($\Delta T = 100$ K), $p = 20$ bar, and $\psi = 0.0$

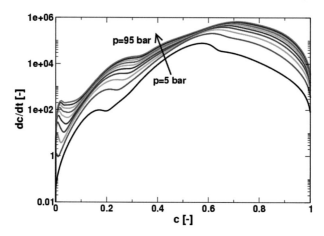

Figure 4.8: Calculated dc/dt profiles as a function of c for $\phi = 1.0$, $T = 800$ K, $5 \leq p \leq 95$ bar ($\Delta p = 10$ bar), and $\psi = 0.0$

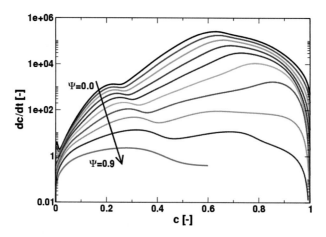

Figure 4.9: Calculated dc/dt profiles as a function of c for $\phi = 1.0$, $T = 800$ K, $p = 20$ bar and $0.0 \leq \psi \leq 0.9$ ($\Delta\psi = 0.1$)

An increase of EGR as displayed in figure 4.9 leads to a continuous decrease of the dc/dt values. While the onsets of the first and second ignition stages are moved towards smaller c values, the main heat release is shifted to higher values of c. For $\psi = 0.9$ the reactions even break off due to the strongly delayed onset of the main heat release. However, the dc/dt profiles of the different EGR rates are not comparable to each other. Thus, a library limitation is not valid in terms of ψ.

Summarising, for library size minimisation a range limitation in ϕ space is valid to $\phi \leq 3.4$. Values greater than $\phi = 3.4$ show a dc/dt profile similar to $\phi = 3.4$. On the fuel lean side, the dc/dt profiles are not comparable, making a limitation not feasible. In T space, a library reduction can be done for $T < 400$ K. This temperature is the lower bound temperature below which no auto-ignition occurs. For high temperatures a library limitation is not feasible in the range examined. The same is true for the full range examined in p and ψ space.

4.5.2   Definition of Independent Variable Nodes

Next to library extension range, the numbers and positions of the independent variables combustion progress, temperature, pressure, fuel-air equivalence ratio and EGR rate need to be defined. Aiming at a certain degree of accuracy, the following investigations build the fundament for the minimum IPV library step size required.

Combustion Progress Variable

The influence of the combustion progress variable nodes on the prediction of ignition delay time is investigated in the following using the homogeneously initialised simplified test case specified in appendix A.1.2. The distributions of the tabulated combustion progress variable nodes in the range $0 \leq c \leq 1$ examined are listed in table 4.2.

Figures 4.10 and 4.11 compare the calculated ignition delay times in terms of temperature increase on the basis of tabulated combustion progress variables with the 0D detailed chemistry solution.

An equidistant distribution of the combustion progress variable nodes with a step size of $\Delta c_{Tab} = 0.01$ underestimates the ignition delay time by $t = 0.001$ s. The difference between the 0D detailed chemistry solution and the calculated solution on the basis of tabulated combustion progress variables reduces heavily with increasing step size up to $t = 0.0002$ s using a step size of $\Delta c_{Tab} = 0.00001$. However, the difference of predicted ignition delay time between a library consisting of 1001 nodes ($\Delta c_{Tab} = 0.001$) and one of 100001 nodes ($\Delta c_{Tab} = 0.00001$) is in the range of $t = 0.0002$ s, too. Hence, a trade-off between combustion progress variable step size and accuracy enhancement can be asserted.

Δc_{Tab} [-]	Number of nodes
Equidistant: 0.01	101
Equidistant: 0.001	1001
Equidistant: 0.0001	10001
Equidistant: 0.00001	100001
Sectional equidistant:	5096
for $c \leq 0.05$: 0.00001, for $c > 0.05$: 0.01	
Logarithmic	20

Table 4.2: Investigated distribution of combustion progress variable nodes

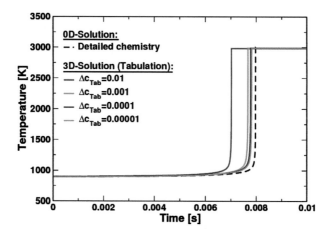

Figure 4.10: Ignition delay times in terms of temperature increase calculated on the basis of tabulated combustion progress variables in equidistant distribution and 0D detailed chemistry solution for $\phi = 1.0$, $T = 900$ K, $p = 20$ bar and $\psi = 0$

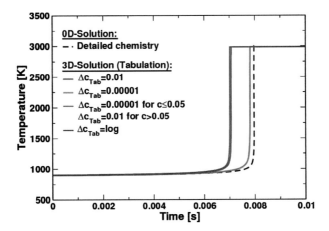

Figure 4.11: Ignition delay times in terms of temperature increase calculated on the basis of tabulated combustion progress variables in sectional equidistant and logarithmic distribution and 0D detailed chemistry solution for $\phi = 1.0$, $T = 900$ K, $p = 20$ bar and $\psi = 0$

Figure 4.11 shows that the strongly reduced step size of $\Delta c_{Tab} = 0.00001$ can be limited to $c \leq 0.05$, indicating the great impact of low combustion progress variables on auto-ignition process. The logarithmic combustion progress variable step size distribution emphasises latter finding. The predicted ignition delay time matches the one of the equidistant $\Delta c_{Tab} = 0.01$ step size distribution, while the in comparison decreased node number of 20 maps primarily low combustion progress variable values.

Summarising, the difference between the direct numerically calculated ignition delay time and the data obtained with the IPV model decreases with increasing c mesh refinement. Thereby, the accuracy enhancement possesses a trade-off to the progress variable step size. Furthermore, the mesh refinement can be limited to combustion progress variables $c \leq 0.05$, indicating the great impact of low c values on auto-ignition. Last finding is emphasised by a logarithmic tabulation of c, which matches the ignition delay time obtained with a course c step size of $\Delta c_{Tab} = 0.01$.

Mixture Defined Independent Variables

In order to investigate the influence of the nodes of the mixture defined independent variables on predicted auto-ignition delay, calculations are carried out using the HCCI engine test case. Details about the engine specifications can be found in A.5.1. The impact of the spark plug initialised flame is excluded.

The basis of the investigations constitutes an IPV library covering the range between $300 \leq T \leq 1200$ K, $5 \leq p \leq 100$ bar[1], $0.2 \leq \phi \leq 4.0$ and $0.1 \leq \psi \leq 0.9$. The independent variables of this basic library are tabulated in an equidistant step size of $\Delta\phi_{Tab} = 0.2$, $\Delta T_{Tab} = 50$ K, $\Delta p_{Tab} = 5$ bar and $\Delta\psi_{Tab} = 0.1$. In the following, the step sizes of the independent variables are varied selectively, as shown in table 4.3.

Variation	$\Delta\phi_{Tab}$ [-]	ΔT_{Tab} [K]	Δp_{Tab} [bar]	$\Delta\psi_{Tab}$ [-]
ϕ	0.1/0.2/0.3/0.4	50	5	0.1
T	0.2	20/50/100	5	0.1
p	0.2	50	5/10/20	0.1
ψ	0.2	50	5	0.1/0.2/0.3

Table 4.3: Investigated selective variations of step sizes of independent variable nodes

As mentioned above, the thermodynamic data can either be tabulated in terms of mean molecular weight M_W and NASA-polynomial coefficients A_i, or specific heat capacity c_p and temperature T. Since these data are tabulated as a function of independent variables, a change in tabulation step size of the independent variables compulsorily has an impact on thermodynamic state calculated. Thus, the investigations are carried out for both types of thermodynamic data retrieval.

Figures 4.12, 4.13, 4.14 and 4.15 depict the calculated in-cylinder pressure curves for the selective variations of the tabulated independent variables step sizes.

---

[1]Note that, to cover the full thermodynamic state range in SI engines, the lower pressure value must be extended to $p = 1$ bar. The results obtained limiting the IPV library to $p \geq 5$ bar are thus not comparable to measurement data.

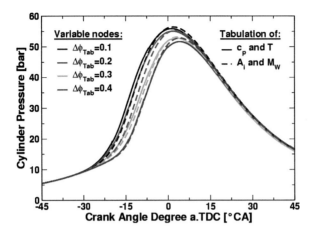

Figure 4.12: Calculated in-cylinder pressures for varying tabulation step sizes of fuel-air equivalence ratio using thermodynamic data tabulated in terms of M_W and A_i as well as c_p and T in HCCI engine

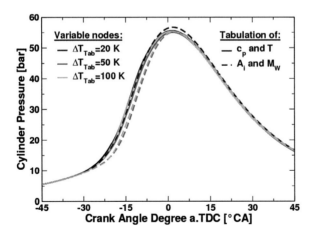

Figure 4.13: Calculated in-cylinder pressures for varying tabulation step sizes of temperature using thermodynamic data tabulated in terms of M_W and A_i as well as c_p and T in HCCI engine

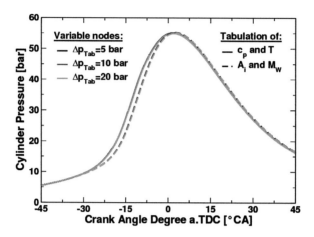

Figure 4.14: Calculated in-cylinder pressures for varying tabulation step sizes of pressure using thermodynamic data tabulated in terms of M_W and A_i as well as c_p and T in HCCI engine

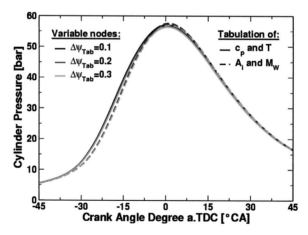

Figure 4.15: Calculated in-cylinder pressures for varying tabulation step sizes of EGR using thermodynamic data tabulated in terms of M_W and A_i as well as c_p and T in HCCI engine

As shown in figure 4.12, a decrease of the step size of the tabulated ϕ results in an almost linear shift of the detected auto-ignition time towards earlier crank angle degrees for both types of thermodynamic data retrieval. The calculated maximum in-cylinder pressure ranges between $p = 5$ bar for the tabulation of A_i and M_W and $p = 4$ bar for the tabulation of c_p and T. Note that, even in case of similar auto-ignition delay time prediction and pressure raise ($\Delta\phi_{Tab} = 0.4$), the maximum in-cylinder pressure differs in a range of $p = 2$ bar using the different types of thermodynamic data tabulation.

Increasing the step size of tabulated temperature values (figure 4.13) has an impact on calculated in-cylinder pressure which is negligible in case of thermodynamic data tabulation in form of c_p and T. In case of tabulation of A_i and M_W the calculated auto-ignition delay time increases from $\Delta T_{Tab} = 20$ K to greater step sizes, whereat the step sizes $\Delta T_{Tab} = 50$ K and $\Delta T_{Tab} = 100$ K show similar delay times. The difference in maximum in-cylinder pressure is in the range of $p = 2$ bar in this case.

A variation of pressure step size possesses qualitatively comparable results for the two ways of thermodynamic data retrieval, as depicted in figure 4.14. Both types of thermodynamic data tabulation detect equal delay times for the step sizes $\Delta p_{Tab} = 5/10/20$ bar.

Finally figure 4.15 shows how varying the step size of EGR ($\Delta\psi_{Tab} = 0.1/0.2/0.3$) the detected auto-ignition delay times remain comparable for thermodynamic data retrieval in form of c_p and T as well as A_i and T.

In order to understand the obtained results in detail, the calculated maximum values of dc/dt of the 0D chemistry solution are mapped as a function of the independent variables for the different step sizes investigated in engine relevant range. Between the data provided, linear interpolation is applied as in the 3D-CFD code.

Figure 4.16 depicts the calculated dc/dt values (white points) as a function of tabulated initial temperatures and tabulated fuel-air equivalence ratios.

Considering the most detailed tabulation of ϕ with a step size of $\Delta\phi_{Tab} = 0.1$, the plot reveals a parabolic shaped decrease of dc/dt with decreasing temperature. The maximum values of dc/dt are located close to $\phi = 1.2$. This shape and its maxima are still maintained by tabulating the independent variable ϕ with a step size of $\Delta\phi_{Tab} = 0.2$. However, the interpolation between the data provided leads to a marginal decrease of the global dc/dt values. An increase of the step size to $\Delta\phi_{Tab} = 0.3$ leads to a wrong prediction of the interpolated dc/dt values between

$1.1 < \phi < 1.4$. Globally, the interpolated dc/dt values considerably underestimate the direct calculated values. Both effects are enhanced by tabulating the fuel-air equivalence ratio in a step size of $\Delta\phi_{Tab} = 0.4$.

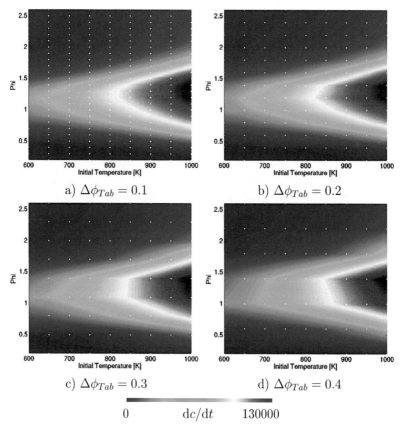

Figure 4.16: Calculated maximum values of dc/dt of 0D chemistry solution (white points) as a function of initial temperature and fuel-air equivalence ratio for different step sizes of fuel-air equivalence ratio tabulated and $p = 20$ bar as well as $\psi = 0$. Contour between the nodes is a result of linear interpolation.

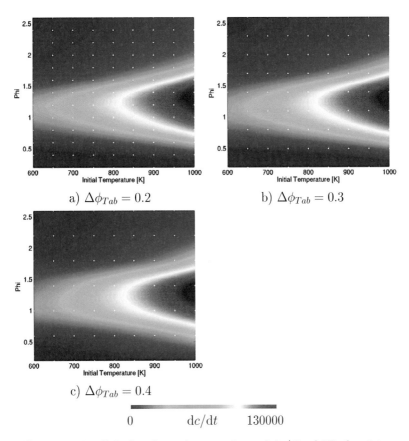

a) $\Delta\phi_{Tab} = 0.2$ b) $\Delta\phi_{Tab} = 0.3$

c) $\Delta\phi_{Tab} = 0.4$

0            $\mathrm{d}c/\mathrm{d}t$ 130000

Figure 4.17: Calculated maximum values of $\mathrm{d}c/\mathrm{d}t$ of 0D chemistry solution (white points) as a function of initial temperature and fuel-air equivalence ratio for different step sizes of fuel-air equivalence ratio tabulated and $p = 20$ bar as well as $\psi = 0$. Contour between the nodes is a result of piecewise cubic Hermite interpolation.

These findings are in-line with the calculated auto-ignition delay times in engine application displayed in figure 4.12. Due to the globally decreased $\mathrm{d}c/\mathrm{d}t$ values, the detected ignition delay time shifts towards higher crank angle degrees with courser ϕ meshes. Concerning the type of thermodynamic data retrieval, figure 4.12 shows that the degree of time shift is less pronounced between $\Delta\phi_{Tab} = 0.1$ and $\Delta\phi_{Tab} = 0.2$ in comparison

to the other step sizes investigated using thermodynamic data tabulated
in form of c_p and T. In comparison, the time shift with decreasing step
size can be described as linear in case of tabulated A_i and M_W. However, the plots in figure 4.16 reveal a decrease of the global dc/dt values
with increasing step size comparable to the detected auto-ignition delay
times using thermodynamic data tabulated in form of c_p and T. Thus,
the tabulation of A_i and M_W involves higher interpolation errors than
the direct tabulation of c_p and T applying linear interpolation. Whereat,
particularly the interpolation of the small numerical values of the fourth
and fifth NASA-polynomial coefficients induces these errors. A decrease
of the interpolation error of tabulated data in higher step sizes demands
for higher order interpolation procedure.

The interpolation errors occurring with courser ϕ meshes can be
reduced applying a higher order interpolation procedure. The interpolation results applying third order piecewise cubic Hermite interpolation
are illustrated in figure 4.17.

In contrast to the linear interpolation procedure the dc/dt characteristic evaluated using cubic interpolation matches the most detailed
tabulation step size of $\Delta\phi_{Tab} = 0.1$ for $\Delta\phi_{Tab} = 0.2$ as well as $\Delta\phi_{Tab} =$
0.3. Using the 3rd order interpolation procedure, the dc/dt map of
fuel-air equivalence ratio tabulated in a step size of $\Delta\phi_{Tab} = 0.4$ is
even comparable to the most detailed data retrieval in a step size of
$\Delta\phi_{Tab} = 0.1$. However, interpolation errors can be observed in the
peripheral regions here. Thus, the cubic interpolation procedure leads
to correct dc/dt characteristic by tabulating ϕ in courser meshes up to
$\Delta\phi_{Tab} = 0.3$. The disadvantage of the 3rd order interpolation procedure
constitutes the high computational demand.

In the following, the different step sizes of tabulated temperatures
are investigated in detail. The calculated dc/dt values and the interpolated data in between are displayed in figure 4.18.

The figure reveals that the parabolic shape of the dc/dt profile in
direction of decreasing temperature with its maximum close to $\phi = 1.2$
is well reproduced independently of temperature tabulation step size investigated. This is in-line with the calculated ignition delay times in
HCCI engine simulation with tabulated c_p and M_W as thermodynamic
data, where the ignition delay times match each other. However, a further decrease of temperature step size would lead to a wrong prediction
of ignition delay times in the range of the fuel-dependent NTC, i.e. between $T = 700, ..., 900$ K.

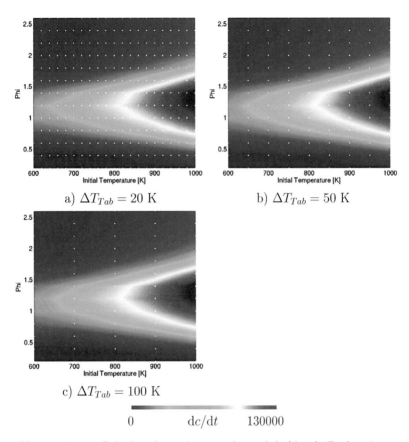

a) $\Delta T_{Tab} = 20$ K

b) $\Delta T_{Tab} = 50$ K

c) $\Delta T_{Tab} = 100$ K

0            $\mathrm{d}c/\mathrm{d}t$ 130000

Figure 4.18: Calculated maximum values of $\mathrm{d}c/\mathrm{d}t$ of 0D chemistry solution (white points) as a function of initial temperature and fuel-air equivalence ratio for different step sizes of temperature tabulated and $p = 20$ bar as well as $\psi = 0$. Contour between the nodes is a result of linear interpolation.

The tabulated dc/dt values and interpolated data of the different step sizes in tabulated pressure values are depicted in figure 4.19.

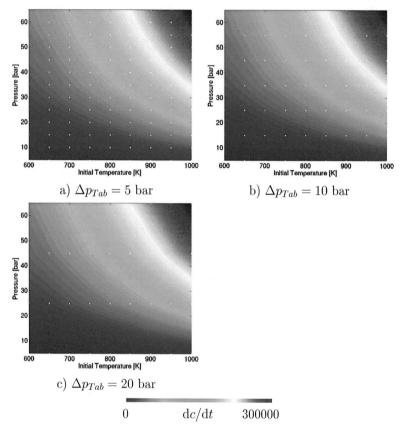

a) $\Delta p_{Tab} = 5$ bar

b) $\Delta p_{Tab} = 10$ bar

c) $\Delta p_{Tab} = 20$ bar

0   dc/dt 300000

Figure 4.19: Calculated maximum values of dc/dt of 0D chemistry solution (white points) as a function of initial temperature and pressure for different step sizes of pressure tabulated and $\phi = 1.0$ as well as $\psi = 0$. Contour between the nodes is a result of linear interpolation.

Tabulating the pressure in a step size of $\Delta p_{Tab} = 5$, 10 or 20 bar has no impact on the interpolated dc/dt values in between the nodes. The interpolated values match each other. This is in-line with the calculated pressure curves in figure 4.14 for both types of thermodynamic data retrieval.

The different step sizes of EGR tabulated are investigated in figure 4.20.

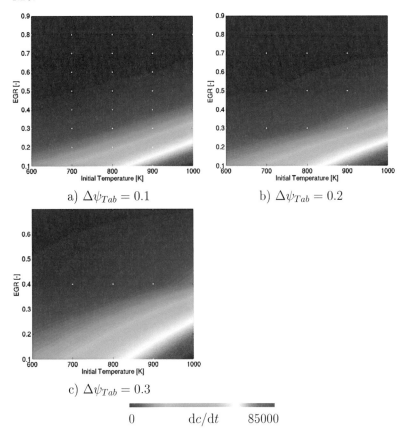

a) $\Delta\psi_{Tab} = 0.1$ b) $\Delta\psi_{Tab} = 0.2$

c) $\Delta\psi_{Tab} = 0.3$

$$0 \qquad \mathrm{d}c/\mathrm{d}t \qquad 85000$$

Figure 4.20: Calculated maximum values of $\mathrm{d}c/\mathrm{d}t$ of 0D chemistry solution (white points) as a function of initial temperature and EGR for different step sizes of EGR tabulated and $\phi = 1.0$ as well as $p = 20$ bar. Contour between the nodes is a result of linear interpolation.

The tabulated step sizes of EGR investigated do not show a difference in interpolated $\mathrm{d}c/\mathrm{d}t$ values. In line with these results, the different tabulated step sizes of EGR possess similar pressure curves in engine simulation for the tabulation of A_i and M_W as well as c_p and T.

Summarising, the tabulation step size of the mixture defined independent variable ϕ has a great impact on auto-ignition time calculated due to interpolation errors of dc/dt occurring. By applying linear interpolation, the interpolation error decreases strongly with increasing ϕ mesh refinement. In order to avoid interpolation errors, a minimal step size of $\Delta\phi = 0.1$ is required. A higher order interpolation procedure enables the tabulation of ϕ in a courser mesh. However, the computational demand in 3D-CFD calculation increases thereby strongly. For the mesh refinement investigated, the tabulated step sizes of the mixture defined independent variables T, p, and ψ do not have an impact on auto-ignition time calculated. Here, the mesh sizes can be decreased to $\Delta T_{Tab} = 100$ K, $\Delta p_{Tab} = 20$ bar, and $\Delta\psi_{Tab} = 0.3$. Concerning the type of thermodynamic data retrieval, the tabulation of c_p and T shows less interpolation errors than A_i and M_W.

4.6   Application for HCCI Engine

In the following, the functionality and interaction of the coupled auto-ignition and flame propagation model is investigated by means of a HCCI engine. The HCCI engine investigated is the Volkswagen Gasoline Compression Ignition (GCI) engine operating in the part load range on the basis of homogeneous auto-ignition in combination with spark ignition. In order to initiate auto-ignition, a high temperature level is initiated in the end of the compression stroke by retracting exhaust gases from the outlet with a systematically repeated opening of the exhaust valves during the intake phase [187]. Details about the retracting strategy and engine specification can be found in appendix A.5.1. The GCI engine characteristic operating range is displayed in figure 4.21.

Based on the test bed measurements three steady-state ran operating points are considered in numerical investigations, which are depicted in figure 4.21 and are listed in table 4.4. Further details about the operating points can be found in appendix A.5.2.

For numerical investigation, the G-equation model is used together with the turbulent flame speed formulation D introduced in section 3.7.1 along with the laminar flame speed fitting function of Perlman. The IPV library provided covers the range $0.2 \leq \phi \leq 3.6$, $400 \leq T \leq 1200$ K, $1 \leq p \leq 100$ bar, and $0.1 \leq \psi \leq 0.9$. The independent variables of the library are tabulated in an equidistant step size of $\Delta\phi_{Tab} = 0.1$, $\Delta T_{Tab} = 100$ K, $\Delta p_{Tab} = 10$ bar, $\Delta\psi_{Tab} = 0.1$, and $\Delta c_{Tab} = 0.01$.

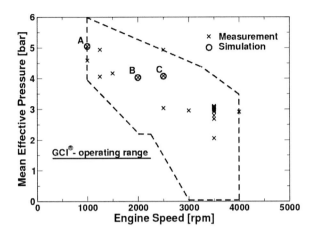

Figure 4.21: Characteristic engine map of the 2.0l GCI engine according to [187]

Operating Point	Engine Speed [rpm]	BMEP [bar]	ψ [-]	$\sigma_{p_{max}}$ [-]
A	1000	5.055	0.24	3.49
B	2000	4.031	0.38	2.78
C	2500	4.082	0.36	2.78

Table 4.4: Investigated operating points of the HCCI engine

Figure 4.22 depicts the measured in-cylinder pressure curves of 200 consecutive cycles (grey dots) for the operating points A to C. Furthermore, the averaged pressure curves of the 200 measured cycles (black dashed lines), which form the boundary conditions for the 3D-CFD simulation, and the calculated curves (black lines) are displayed.

At low engine speeds (operating point A, figure 4.22 a) the coupled auto-ignition and flame propagation model overestimates the averaged in-cylinder pressure. The conditions constituting the auto-ignition process occur early before TDC. As a result, the pressure increases with a heavy gradient leading to high maximum in-cylinder pressure value. However, the pressure curve calculated represents an Indicated Mean Effective Pressure (IMEP) of 6.205 bar, which matches the theoretical value[1] of $6.0 \leq$ IMEP ≤ 6.3 bar. In contrast, the averaged in-cylinder pressure curve describes an IMEP of 5.437 bar. This can be assigned to the high standard deviation of $\sigma_{p_{max}} = 3.49$. The pressure curves of the 200 consecutive cycles show a wide spread of the maximal in-cylinder pressure values, including a high number of spark plug and auto-ignition failures as well as retarded combustion[2] (lower maximal in-cylinder pressure values). This spread can be ascribed to the low engine speed, which provokes mixture non-homogeneities and thus cyclic variations.

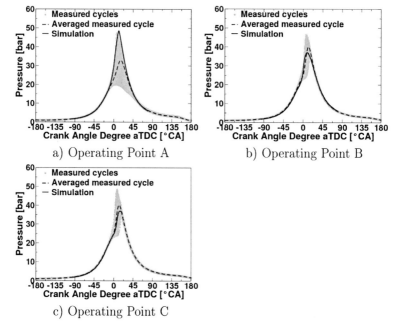

a) Operating Point A   b) Operating Point B

c) Operating Point C

Figure 4.22: Calculated and measured pressure curves for HCCI engine at operating points specified in table 4.4 [16]

With increasing engine speed (operating points B and C, figures 4.22 b) and c) the standard deviation of the measured in-cylinder pressure curves reduces. The decreased deviation leads to a good repro-

---

[1]The IMEP is defined as sum of BMEP and Frictional Mean Effective Pressure (FMEP) [70]. According to Heywood [70] and Golloch [57], in SI engines FMEP is in the range of $1.0 \leq$ FMEP ≤ 1.3 bar.

[2]Details about combustion duration and 50 % transformation point of the measured pressure curves can be found in appendix A.5.3.

duction of the averaged pressure curves of the measurements using the coupled combustion model. Both the calculated onsets of combustion as well as the maximum in-cylinder pressure values match the averaged measurement data.

In the following, the functionality and interaction of the individual combustion models are investigated in detail considering operating point C. The operating point shows a low standard deviation for the test bed measurements and can be reproduced well using the coupled autoignition and flame propagation model. Figure 4.23 displays the averaged pressure curve (black dashed line) of the 200 cycles measured. Additionally, the plot displays the calculated pressure curves resulting using the following combustion models: G-equation model only (red curve), IPV model only (green curve), and coupled G-equation / IPV model (blue curve).

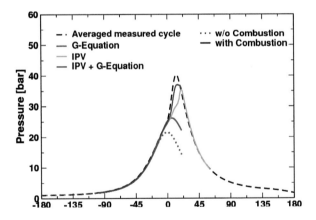

Figure 4.23: Calculated impact of the separate and coupled combustion models on the pressure curve for the HCCI engine at operating point C

In figure 4.24, temperature planes through the combustion chamber along with an iso-surface representing the combustion progress variable $c = 0.5$ (in magenta) are shown for the different model applications.

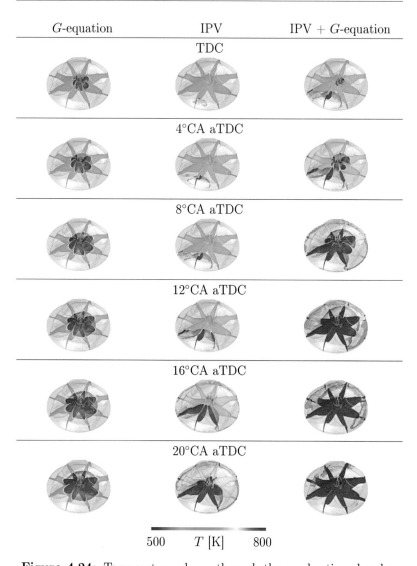

Figure 4.24: Temperature planes through the combustion chamber and iso-surface representing the combustion progress variable $c = 0.5$ (in magenta) for separate and coupled application of combustion models for the HCCI engine at operating point C

Using the G-equation model only, the mixture in the combustion chamber is transformed starting from the spark plug. This corresponds to a conventional SI engine combustion. The flame propagation speed is strongly reduced due to the high amount of residual gas present in the combustion chamber. As a result, the pressure gradient and thus the maximal in-cylinder pressure are low.

Besides the spark plug initiated transformation of the in-cylinder mixture, auto-ignition phenomena occur in HCCI combustion. Applying the auto-ignition model only, the mixture in the combustion chamber is transformed starting from regions satisfying auto-ignition conditions. These conditions are matched at TDC between the intake and exhaust valves due to predominating stoichiometric conditions and low ψ fraction of the mixture. Successively, the first hot spot extends. A second auto-ignition kernel arises at 12°CA aTDC between the intake valves. At 16°CA aTDC a multitude of auto-ignition kernels arise close to the combustion chamber wall in the whole cylinder. As a consequence the pressure and temperature levels increases marginally in the chamber and the individual hot spots start to transform the remaining mixture promptly. The transformation of the in-cylinder charge is dominated by the first hot spot. The prompt transformation of the mixture is also apparent from the pressure curve. Until 16°CA aTDC the pressure increases only marginally. Thereafter, the pressure gradient is enhanced strongly.

In reality, the HCCI combustion is a combination of auto-ignition and flame propagation. The spark plug initialised flame transforms a small amount of the cylinder charge. The figure 4.24 reveals that the modified thermodynamic conditions by the IPV model damp the initial flame propagation (TDC). The temperature and pressure increase as a result of flame propagation leads to an acceleration of auto-ignition processes in the unburnt mixture. The calculated resulting pressure gradient matches the experimental data well.

This investigation shows that the coupled G-equation / IPV modelling approach works well for simulating the auto-ignition process in a HCCI engine. The method allows for examining the interaction between flame propagation and locally appearing auto-ignition processes in the unburnt mixture. However, the method fails to predict an average HCCI engine combustion process for operating points with a high standard deviation in test bed measurements. In order to account for the high cycle deviation, multi-cycle calculations are required.

4.7   Application for Highly Turbocharged SI Engine

The IPV model capability to predict pre-ignition phenomena is investigated in the following by considering a highly turbocharged SI engine. The highly turbocharged SI engine examined has a supercharger and an exhaust gas turbocharger serially connected. At low engine speeds the compressor is used in addition to the exhaust gas turbocharger to enhance the boost pressure. The heavily enhanced boost rates at low engine speeds make the engine feasible especially for the investigation of pre-ignition phenomena [19]. Additionally, at low engine revolution speeds a tumble flap located in the intake port can be activated to enhance the in-cylinder charge motion. The increased charge motion leads to a decrease of cyclic variations and a shortened combustion duration [19]. Further details about the engine specifications can be found in appendix A.2.1.

In the following, the pre-ignition tendency of the highly turbocharged SI engine is investigated at full load operating point (1500 rpm, 20.17 bar) for a tumble flap configuration variation (opened / closed).

Pre-ignition phenomena occur stochastically. The minor reproducibility of this phenomena demands for a multi-cycle approach in 3D-CFD. The procedure of such a multi-cycle calculation is outlined in the following.

The calculation starts 90 °CA bTDC. Due to the low turbulence level present in the combustion chamber in the first calculation cycle, the flame propagation speed is low and only a small amount of the mixture within the chamber is transformed. In the next cycle, the in-cylinder turbulence level reaches adequate values accelerating the flame propagation and thus mixture transformation. The resulting pressure curve is validated with measurement data of an averaged non-pre-ignition cycle. Starting from the following cycle, the IPV model is turned on allowing the prediction of pre-ignition phenomena in the combustion chamber. In total, seven cycles are considered to evaluate the pre-ignition tendency for the same operating strategy. For numerical investigation, the G-equation model is used together with the turbulent flame speed formulation D introduced in section 3.7.1 and the laminar flame speed fitting function of Perlman. The IPV library provided covers the range $0.2 \leq \phi \leq 3.6$, $400 \leq T \leq 1600$ K, $1 \leq p \leq 140$ bar, and $0.0 \leq \psi \leq 0.9$. The independent variables of the library are tabulated in an equidistant step size of $\Delta\phi_{Tab} = 0.1$, $\Delta T_{Tab} = 100$ K, $\Delta p_{Tab} = 10$ bar, $\Delta\psi_{Tab} = 0.1$, and $\Delta c_{Tab} = 0.01$.

Figure 4.25 displays the calculated in-cylinder pressure curves of the seven consecutive cycles of the tumble flap position variation. Note that at gas exchange the in-cylinder pressure trace is interpolated.

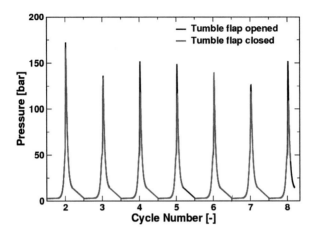

Figure 4.25: Calculated pressure curves of highly turbocharged SI engine at full load operating point (1500 rpm, 20.17 bar) and varying tumble flap position

In the first calculation cycle, the in-cylinder pressures of both tumble flap positions increase up to 175 bar. The pressures of the tumble flap opened and closed positions decrease in the second cycle reaching maximum values of 140 bar. In the third and fourth calculation cycle striking differences between the tumble flap positions become evident. While the in-cylinder pressures of the tumble flap opened position increase up to 150 bar, the pressures of the tumble flap closed position reach maximum values of 120 bar. The fifth and sixth calculation cycles show comparable maximum in-cylinder pressure values of the tumble flap positions, i.e. 140 and 125 bar respectively. In the last cycle differences in the cylinder pressure between the tumble flap positions can be observed again. Here the in-cylinder pressure of the tumble flap opened position increases up to 150 bar and the maximum pressure of the tumble flap closed position is decreased to 110 bar.

In general, the pressure curves reveal an increased auto-ignition tendency of the tumble flap opened configuration in comparison with the tumble flap closed one. The total values of the tumble flap opened position are always greater than the one of the tumble flap closed position, indicating an earlier start of combustion. However, the pressure curves do not reveal whether the pressure increase is due to pre-ignition or engine knock. This can be achieved by considering the timing of five percent transformation $x_b = 0.05$, evaluated from the rate of heat release. The five percent transformation point is calculated for each calculation cycle for both tumble flap configurations. The resulting distribution is displayed in figure 4.26 along with the spark timing, constituting the threshold between pre-ignition and engine knock.

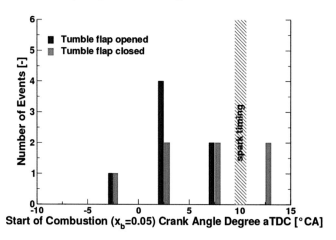

Figure 4.26: Calculated distribution of five percent transformation points classified in time intervals of highly turbocharged engine at full load operating point (1500rpm, 20.17 bar) and tumble flap opened and closed configurations

The plot reveals that all combustion onsets of the tumble flap opened configuration are located before spark timing. Hence, the enhanced in-cylinder pressures displayed in figure 4.25 are a result of pre-ignition processes. In contrast, only five cycles of the tumble flap closed position can be assigned to pre-ignition phenomena.

Considering the distribution of start of combustion before spark timing, four of the total seven cycles of the tumble flap opened position show their combustion onset in the time interval 0 to 5°CA aTDC. Thus, the pre-ignition is initiated in most cases directly after reaching the critical in-cylinder pressure and temperature conditions at TDC.

The combustion onsets of the tumble flap closed configuration are uniformly distributed in the time intervals 0 to 5°CA aTDC as well as 5 to 10°CA aTDC. The comparatively retarded combustion onsets can either be related to extenuated critical in-cylinder conditions at TDC or decelerated reaction progresses.

In order to understand the differences of pre-ignition tendency of the tumble flap variation in detail, the in-cylinder auto-ignition processes of the individual cycles are examined in the following. Figures 4.27 and 4.28 display the in-cylinder combustion progress variable distribution close to TDC of the individual cycles of the tumble flap opened and closed configurations.

The first cycle of the tumble flap opened configuration (figure 4.27) possesses a first combustion progress variable increase located close to the piston surface on the intake valve side at 5°CA bTDC. The auto-ignited area grows continuously and at TDC two additional auto-ignition kernels can be observed. One region is located close to the combustion chamber wall at the intake valve face. The other new auto-ignition kernel arises close to the combustion chamber roof on the exhaust valve side. At 5°CA aTDC the auto-ignited areas cover a large part of the combustion chamber and a number of additional hot spots are formed close to the combustion chamber wall, merging the first auto-ignition areas subsequently.

In the second cycle an initial increase of the combustion progress variable can be observed at 5°CA aTDC. The areas of progress variable increase are equal to the one identified for the first calculation cycle. However, the region on the exhaust side becomes predominant.

Cycle three and four show a remarkable combustion progress variable increase at TDC. The auto-ignition regions match the one of the first cycle, too. The predominant impact on mixture transformation of the hot spot located close to the piston surface on the intake valve side becomes in these two cycles less pronounced.

In the fifth cycle the hot spot located close to the combustion chamber roof on the exhaust valve side turns up first at 5°CA aTDC. The dominating auto-ignition area in the sixth cycle is the one located close to the combustion chamber wall at the intake valve face and occurs first at 5°CA bTDC. In cycle seven the first hot spots occur at TDC. As in cycle five, the auto-ignition area located close to the combustion chamber roof on the exhaust valve side is predominant here, too.

Cycle 1   Cycle 2   Cycle 3   Cycle 4   Cycle 5   Cycle 6   Cycle 7

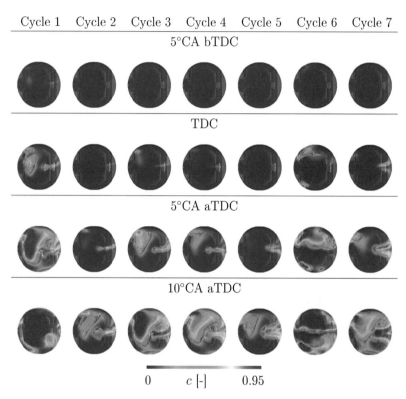

Figure 4.27: Combustion progress variable planes through the combustion chamber and iso-surface of the combustion progress variable $c=0.3$ (magenta) of the highly turbocharged engine at a full load operating point (1500 rpm, 20.17 bar) and tumble flap opened configuration

The first hot spot revealing in the first cycle of the tumble flap closed position (figure 4.28) is located on the exhaust valve side near the squish region and arises at 5°CA bTDC. At TDC a number of additional hot spots arise, which are located on the exhaust valve side, intake valve side close to the combustion chamber wall as well as chamber spout. The auto-ignition areas grow continuously and merge subsequently.

In cycles two to five as well as seven, the first auto-ignition kernel is located on the exhaust valve side near the squish region. In these cycles only the timing of formation differs.

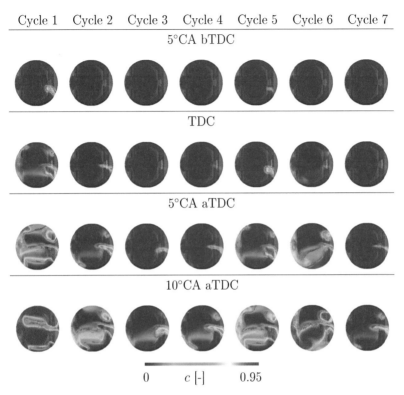

| Cycle 1 | Cycle 2 | Cycle 3 | Cycle 4 | Cycle 5 | Cycle 6 | Cycle 7 |

5°CA bTDC

TDC

5°CA aTDC

10°CA aTDC

0   c [-]   0.95

Figure 4.28: Combustion progress variable planes through the combustion chamber and iso-surface of the combustion progress variable c=0.3 (magenta) of the highly turbocharged engine at a full load operating point (1500 rpm, 20.17 bar) and tumble flap closed configuration

An exception occurs in the sixth cycle. Here hot spots appear first at TDC and are located on the exhaust as well as intake valve side.

The results obtained match qualitatively the optical measurements achieved using a waveguide accessing the combustion chamber via spark plug. The optical waveguide consists of 40 channels arranged in one layer, allowing the detection of light emission along the combustion chamber roof top to chamber wall. The distribution of measured initial auto-ignition origin locations over the individual channels of 200 successive cycles is displayed in figure 4.29.

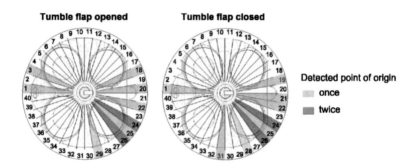

Figure 4.29: Measured auto-ignition origin locations of the tumble flap variation using an optical 40-channel waveguide [18]

The optical measurements show that most of pre-ignition phenomena originate on the exhaust valve side for both tumble flap configurations. In contrast, the number of initial auto-ignitions detected on the intake valve side is comparatively small. Herein, the tumble flap opened position shows a higher tendency to auto-ignite on the intake valve side than the tumble flap closed configuration.

Cycle 1	Cycle 2	Cycle 3	Cycle 4	Cycle 5	Cycle 6	Cycle 7

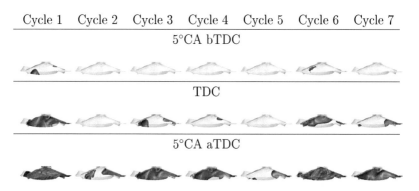

Figure 4.30: Iso-surface of combustion progress variable $c \geq 0.1$ (magenta) of the highly turbocharged engine at a full load operating point (1500 rpm, 20.17 bar) and tumble flap opened position

The calculated pre-ignition locations of the tumble flap closed position match the measurement data well. The auto-ignition locations of the tumble flap opened position predicted in 3D-CFD simulation show two major origins. One is located at the combustion chamber roof on the exhaust valve side and the other one close to the piston surface on the intake valve side. Thus, the results are not in-line with the measurements. However, figure 4.30 shows how hot spots located close to the piston surface can not be predicted using an optical waveguide arranged in one layer along the combustion chamber roof. Taking into account, that the two major auto-ignition kernels detected in the simulation occur in average at the same time, it becomes evident that the optical results predominantly detect the auto-ignition kernel on the exhaust valve side.

In order to examine the auto-ignition process of the tumble flap configurations, the variables influencing the ignition process are investigated in detail in the following. Figures 4.31 and 4.32 depict the in-cylinder distribution of the reaction progress variable, temperature, fuel-air equivalence ratio, and EGR in the time interval 40°CA bTDC until 5°CA bTDC for both tumble flap configurations.

At 40 °CA bTDC for the tumble flap opened position the temperature varies in a range of 50 K. High values can be observed close to the chamber wall between the intake and exhaust side as well as the squish region. The increased temperature regions between the ports are characterised by a low fuel-air equivalence ratio and high EGR rates. In the squish region, only a marginal amount of EGR is present and the fuel-air equivalence ratio equals one. The region close to the injector on the intake valve side shows comparatively low temperature values primarily due to fuel rich conditions.

In the next crank angle degrees until 20 °CA bTDC the mixture distribution is only marginally homogeneoused. Thus, the characteristics are comparable to the one at 40 °CA bTDC. At the same time, auto-ignition areas begin to develop. The auto-ignition kernel close to the piston surface on the intake valve side originates in an area characterised by a relatively high temperature and fuel-air equivalence ratio and low EGR rate. The same characteristic is true for the hot spot originating close to the combustion chamber roof on the exhaust valve side.

The auto-ignition areas grow successively primarily into regions characterised by an increased temperature and fuel-air equivalence ratio as well as low EGR rate.

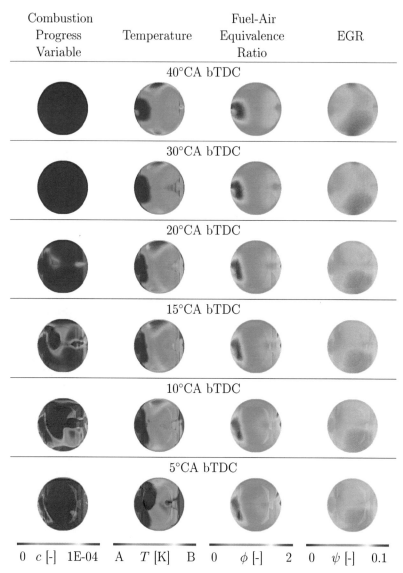

Figure 4.31: Variable planes through the combustion chamber in cycle one of tumble flap opened position of the highly turbocharged SI engine at a full load operating point (1500 rpm, 20.17 bar). The minimal and maximal temperature values are specified in table 4.5.

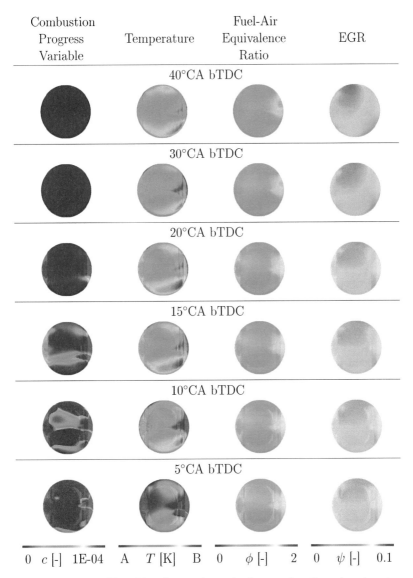

Figure 4.32: Variable planes through the combustion chamber in cycle one of tumble flap closed position of the highly turbocharged SI engine at a full load operating point (1500 rpm, 20.17 bar). The minimal and maximal temperature values are specified in table 4.5.

Degree bTDC	40°CA	30°CA	20°CA	15°CA	10°CA	5°CA
A [K]	550	600	640	660	680	700
B [K]	600	670	690	710	730	750

Table 4.5: Minimal and maximal temperature values set in figures 4.31 and 4.32

In contrast to the tumble flap opened position, the in-cylinder distribution of the tumble flap closed position can be described as homogeneous. At 40 °CA bTDC relatively high temperature fields can be observed primarily on the exhaust valve side between the ports due to decreased fuel-air equivalence ratio. The squish region is characterised by low temperature values and high fuel-air equivalence ratio.

As for the tumble flap opened position, the mixture characteristic at 20 °CA bTDC is comparable to the one at 40 °CA bTDC. At this time, a first progress variable increase can be observed in the squish region, where high temperature and fuel-air equivalence ratio as well as low EGR rates are present. 5 °CA later, a second hot spot arises on the exhaust valve side. Both auto-ignition areas extend into regions characterised by favourable conditions for auto-ignition.

Summarising, pre-ignition phenomena are a result of mixture non-homogeneity. High fuel-air equivalence ratio and temperature fields enhance the ignition process, thus being the major source of pre-ignition phenomena. The tumble flap closed configuration possesses in comparison with the tumble flap opened position an increase of in-cylinder mixture homogeneity. As a result, the auto-ignition tendency is reduced. The second source of pre-ignition phenomena are increased temperatures on the exhaust valves.

The investigation shows how the IPV model allows for the calculation of auto-ignition phenomena in terms of pre-ignition. Thereby, the correct prediction of auto-ignition tendency demands for multi-cycle calculations.

Chapter 5

Emission Formation Modelling

5.1 Introduction

Combustion processes lead, besides of the formation of stoichiometric combustion products, to the formation of air pollutants. These are primarily carbon monoxide (CO), unburnt Hydro-Carbons (HCs), nitrogen monoxide (NO), and soot. The formation of these emissions is dependent on fuel type and boundary conditions. For *iso*-Octane fuel, the dependency of emission formation on temperature and fuel-air equivalence ratio is depicted in figure 5.1.

The figure reveals, that the CO formation (figure 5.1 a) is primarily controlled by the fuel-air equivalence ratio. In absence of O_2, the CO oxidation reaction $CO + OH \leftrightarrows CO_2 + H$ competes against $H_2 + OH \leftrightarrows H_2O + H$. the last reaction is in partial equilibrium, while the CO oxidation reaction is strongly kinetically controlled [118]. Thus, the CO concentration increases with increasing fuel-air equivalence ratio.

Moreover, figure 5.1 b shows, that unburnt hydrocarbons usually result from rich mixture zones at low temperatures. HCs are thus the consequence of incomplete combustion of fuel. However, over-leaning is also an important source of HC formation [118].

The formation of NO (figure 5.1 c) is primarily controlled by the temperature. The so-called *thermal* NO is formed from atmospheric nitrogen at high temperatures [194]. NO formation processes possessing a reaction rate which is greater than the one of the thermal NO formation path are called *prompt* NO mechanism [120].

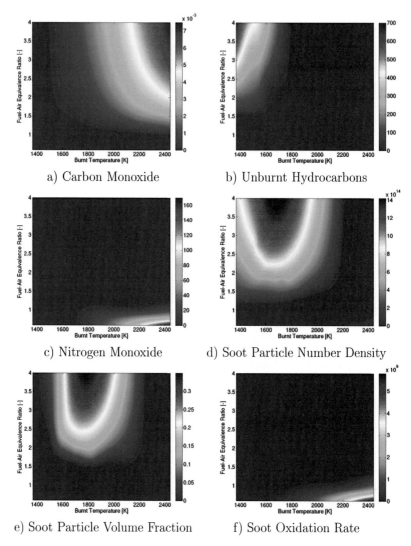

a) Carbon Monoxide   b) Unburnt Hydrocarbons

c) Nitrogen Monoxide   d) Soot Particle Number Density

e) Soot Particle Volume Fraction   f) Soot Oxidation Rate

Figure 5.1: Integrated mole fractions of CO, HC, NO, soot particle number density (in $[1/\mathrm{cm}^3]$), soot volume fraction (in $[\mathrm{m}^3/\mathrm{m}^3]$), and soot oxidation rate (in $[1/\mathrm{s}]$) calculated in 0D homogeneous reactor calculations for $p = 40$ bar and $\psi = 0.0$

The prompt NO mechanism are: a) an increased NO production due to *super-equilibrium* of O and OH, which occurs typically in the flame front, b) *Fenimore* NO, which is formed in flame regions possessing an excess of radical CH, c) a trimolecular recombination leading to *nitrous* NO formation, which occurs typically under high pressure and fuel lean conditions, and d) *fuel bounded* NO. In general, the NO formation increases with temperature and decreasing fuel-air equivalence ratio.

The soot formation process involves a multitude of chemical as well as physical processes[1], which depend on pressure, mixture composition, temperature as well as fuel type [113]. These processes incorporate the pyrolysis of the fuel, particle inception, coagulation of particles to larger aggregates, and heterogeneous surface reactions of the soot particles with gas phase species, leading to particle growth or oxidation. The size distribution of the soot particles formed ranges from particles containing less than hundred up to particles containing a few million of carbon atoms [125].

Figure 5.1 d reveals, that soot is formed in regions where the fuel-air equivalence ratio is high and the temperature high enough to promote pyrolysis of fuel. Coagulation and heterogeneous surface reaction processes lead to an increase of the soot volume at higher temperatures (figure 5.1 e). However, a high temperature and pressure level leads in general to an increase of the soot formation, but also to an increase of soot oxidation processes occurring under fuel lean conditions [79], which reduces the overall amount of particles (figure 5.1 f).

In conventional diesel engines fuel and air are mixed by convection and diffusion during combustion, and the auto-ignition based mixture transformation leads to high in-cylinder temperature and pressure values. These conditions promote primarily the formation of NO and soot. Whereas, conventional SI engines operate under premixed conditions. Applying port fuel injection, the premixed flame initiated mixture transformation leads, in comparison with diesel engines, to lower in-cylinder temperature and pressure values. Under these conditions, primarily CO and HC can be formed. However, modern turbocharged SI engines with direct injection possess much higher in-cylinder temperature and pressure values. Moreover, applying direct injection, the time for mixture formation is decreased, and spray impingement can lead to the formation of wall film. In consequence, the formation processes of HC and soot are promoted.

---

[1]The processes involved in soot formation process are still contended [52, 83]. An overview of the general accepted as well as contended processes involved in the soot formation are found in [73, 110].

In order to model the emission formation process, different modelling approaches are proposed in the literature. A brief survey of these models is given in section 5.2. In this work, an interactively coupled flamelet model is used for the modelling of premixed SI engine emission formation. The model is introduced in section 5.3. The description of the soot formation bases on a detailed soot model. Instead of solving the different soot particle size classes in a multitude of partial differential equations, the properties of the soot Particle Size Distribution Function (PSDF) are calculated using the method of moments. The soot model and moment closure are described in section 5.4. The coupling of the flamelet model to the 3D-CFD code is shown in section 5.5. The flamelet model is used to model the emission formation in burning and burnt zone regions. This demands for a flamelet initialisation as partially burnt. By considering the combustion as premixed and thus the scalar dissipation rate as low, the flamelet model bases on an operator splitting procedure, enabling the flamelet initialisation by tabulated 0D homogeneous reactor solutions. This procedure and resulting differences to full flamelet solutions are shown in section 5.6. The emissions calculated with the interactively coupled flamelet model are compared with 0D homogeneous reactor calculations in section 5.7. The flamelet model bases on a single flamelet modelling approach. The functionality of this approach to model the emission formation in separated burning volumes is investigated in section 5.8.1. The validity of the premixed flame and low scalar dissipation rate assumption is examined in section 5.8.2. Afterwards, the model functionality to calculate emission formation in SI engines is investigated on a Start Of Injection (SOI) variation of an optical SI engine in section 5.9.

5.2   Literature Survey

Besides aspects of chemistry and physics, research in emission formation modelling is often motivated from technical devices such as diesel engines or gas turbines [116]. In these devices combustion occurs in turbulent diffusion flames. Emission formation modelling approaches described in the literature for premixed combustion processes, as they occur in SI engines, confine on feasibility studies, like the work of Dekena [43]. The literature survey on 3D-CFD emission formation modelling is thus restrained to diesel engine application in the following.

The prediction of fuel specific emission formation demands for a model describing the overall combustion and subsequent pollutant formation process. The approaches described in the literature can be classified - in analogy to the auto-ignition modelling approaches (see section 4.2) - into models incorporating detailed chemistry and models which do not.

Approaches which do not account for detailed chemistry base on a global reaction step incorporating five or seven species [70] and a characteristic turbulent time assuming that the combustion is controlled primarily by the mixing process [12]. Using these kind of combustion models, the emission formation needs to be modelled separately. In diesel engine applications [91], typically additional sub-models for the NO and soot formation are applied. The soot models are of empirical or semi-empirical nature[12] [72, 82], like the Hiroyasu soot formation model [71, 128], which is often used in combination with the Nagle-Strickland-Constable oxidation model [123]. Due to the limited number of species involved and subsequent separated combustion and emission formation modelling, the models capability to predict engine processes is highly restricted and demands for significant model parameter tuning [12].

Emission formation modelling approaches incorporating detailed chemistry can either base on a direct integration of the chemistry or a decoupled solution of the flow field and chemistry.

The direct integration of the chemistry involves the coupling of the wide range of flow variables with the even smaller chemical time scales. The solution of the small chemical time scales increases the computational demand strongly.

A decoupling of the detailed chemistry emission formation calculation from the flow field modelling can be done based on the laminar flamelet concept [133]. The laminar flamelet concept covers regimes of premixed and non-premixed combustion and bases on the flamelet assumption (see 3.5) [133]. This assumption enables the transformation of the species mass balance and enthalpy conservation equations and the calculation of the inner structure of the laminar reaction zone in an one-dimensional form. Thus, the numerical effort incorporated with the solution of the small chemical time scales can be separated from the solution of the flow field.

Especially the soot formation process is sensitive to flame temperature, radical levels, and pressure [80]. Thus, the demand for a modelling based on detailed chemistry is much higher.

---

[1]An extensive overview of empirical and semi-empirical soot models can be found in [83].

[2]Coupling the 3D-CFD combustion simulation with an emission model in an indirect way disables the application of a detailed kinetic soot model [58, 59].

Since the size of a soot particle influences its characteristic, different size classes of the soot need to be modelled [125]. This results in an infinite number of partial differential equations. Instead of solving the infinite number of equations, mathematical models are used, describing the development of the PSDF [126]. Conservation equations for the properties of the PSDF are formulated using the method of moments [51, 52] and the sectional method [55]. While the method of moments provides information on statistical quantities of the PSDF, the sectional method provides the actual shape of the PSDF by dividing the size distribution functions into a discrete number of sections [126]. In the sectional method, the transport property is a density weighted soot volume fraction for all particles within a section [116]. However, the computational demand for the solution of the density weighted soot volume fraction transport equations in 3D-CFD is nowadays still too high. The transport property of the method of moments is the r-th moment of the PSDF. The computational demand for the solution of the moment transport equations is relatively small, which makes this approach to an appealing and widely used [80, 121, 124, 138, 180, 183] solution for the calculation of soot formation in 3D-CFD.

5.3   TIF Model Description

The flamelet approach bases on the assumption, that species mass fractions and the enthalpy are a function of one conserved, normalised scalar - the mixture fraction Z [13], which is related to the fuel-air equivalence ratio ϕ according to equation 3.35. In order to express the scalar variables as a function of Z, a locally defined coordinate system is introduced. The coordinate $x_1 = Z$ of this coordinate system is perpendicular to the flamelet iso-surface, while x_2 and x_3 are within this iso-surface. Using a Crocco type transformation, the species and enthalpy conservation equations can be translated into one-dimensional form, thus enabling the computation of the inner flame structure [134].

The species and enthalpy conservation equations obtained through the transformation read under the assumption of unity Lewis number [133]

$$\rho \frac{\partial Y_i}{\partial t} - \rho \frac{\chi}{2} \frac{\partial^2 Y_i}{\partial Z^2} = \omega_i \tag{5.1}$$

$$\rho \frac{\partial H}{\partial t} - \rho \frac{\chi}{2} \frac{\partial^2 H}{\partial Z^2} = \frac{\partial p}{\partial t} + q_{rad}. \tag{5.2}$$

In equation 5.1, the source term ω_i accounts for formation and consumption of mass fraction Y of species i due to chemical reactions and is calculated according to equation 3.24. In equation 5.2, the term $\partial p / \partial t$ becomes zero by assuming a quasi-constant pressure[1]. In the following, radiation losses q_{rad} are neglected.

In consequence of the transformation in mixture fraction space, the conservation equations 5.1 and 5.2 possess no convective terms [13]. As a result, no relative convective velocities exist between the mixture fraction and other scalars. The impact of the flow field is represented by the scalar dissipation rate only,

$$\chi = 2D_Z \left(\nabla Z \right)^2 \tag{5.3}$$

which represents instantaneous local diffusion and strain effects by the flow field [13], with D_Z as diffusion coefficient.

In engine applications, the mixture fraction and therefore the scalar dissipation rate vary spatially and in time. Under varying boundary conditions, steady-state flamelet modelling is not valid [115]. In order to model the impact of χ under varying and representative boundary conditions, unsteady flamelet modelling is required. Thus, the flamelet needs to be coupled to the 3D-CFD flow field interactively.

5.4 Detailed Soot Model Description

5.4.1 Detailed Kinetic Soot Model

In this work, the detailed kinetic soot model of Mauß [114] is used to model the emission formation in SI engine combustion processes. This model bases on Frenklach's soot model [50] and provides a detailed chemical and physical description of soot formation and oxidation processes. The soot formation and oxidation processes involve gas phase reactions, polymerisation of Polycyclic Aromatic Hydrocarbons (PAHs), formation and growth of soot particles due to processes involving particle inception, surface growth, PAH condensation, particle coagulation, and soot particle oxidation [80].

In *gas phase reactions*, the hydrocarbon fuel pyrolysis and oxidation process leads to the formation of aromatic species. These aromatic species are either intermediate species, like benzene, or directly abstracted from the fuel. The aromatic species react further with acetylene and grow forming polyaromatic structures. The largest polyaro-

---

[1]In most low Mach number applications the pressure can be assumed to be spatially constant [13].

matic structure[1] formed constitutes the interface between the gas phase
and the PAH polymerisation process. Based on the largest gas phase
mol-ecule, the PAH species grow further in the *PAH polymerisation*
process by repeated cycles of Hydrogen-Abstraction and C_2H_2-Addition
(HACA) [51]. Thereby, the addition of two acetylene molecules results
in a ring closure and a new six-membered aromatic ring is added. In
contrast to Frenklach's approach [51], latter process is modelled as a
fast polymerisation process[2]. This procedure provides a system of algebraic equations for the concentrations of PAH molecules [113]. The
polymerised PAH molecules collide and stick to each other forming primary soot particles[3] (*particle inception*) [113]. Furthermore, the PAH
molecules collide with soot particles of different size classes and coalesce
with them (*PAH condensation*). In turn, the soot particles formed collide with other soot particles and coalesce (*particle coagulation*). Heterogeneous surface reactions with gas phase species lead to *surface growth*
of the soot particles or *oxidation* of the particles.

The underlying physical soot formation processes base on collision
phenomena due to size dependent motion of the particles, and are modelled using Smoluchowski's equation [116, 125]

$$\frac{dN_i}{dt} = \frac{1}{2} \sum_{j=1}^{i-1} \left(\beta_{j,i-j} N_j N_{i-j} \right) - \sum_{j=1}^{\infty} \left(\beta_{i,j} N_i N_j \right). \qquad (5.4)$$

Thereby, the change rate of the number of particles N of size i is determined by the collision of two particles forming a new particle of size
i (first term on the right hand side of equation 5.4) and the collision
of particles of size i with other particles (second term on the right
hand side). The collision frequency factor $\beta_{i,j}$ is primarily dependent
on the temperature and size of colliding particles and reads in general
form [114, 125]:

$$\beta_{i,j} = \epsilon_{i,j} \sqrt{\frac{8 \pi k_B T}{\mu_{i,j}}} \left(r_i + r_j \right)^2 \qquad (5.5)$$

In equation 5.5, $\epsilon_{i,j}$ denotes the size dependent coagulation enhancement
factor due to attractive forces, k_B the Boltzmann constant, $\mu_{i,j}$ the

---

[1]The largest polyaromatic species of a gas phase reaction mechanism is typically a 4-ring
aromatic species.

[2]Under the assumption, that PAH reactions are fast and accordingly the formation of
soot particles from PAH is fast, the PAH molecules can be considered in steady-state [113].

[3]In case *gas phase* PAH molecules collide to three-dimensional structures which stick
due to Van-der-Waals forces, a primary *solid* soot particle is formed [125].

reduced mass, and r_i the radius of particles of size i. With $V_i = m_i/\rho_s = i\,m_1/\rho_s$, where m_1 being the mass of the smallest size unit, which is assumed to be two carbon atoms[1], and V_i being the volume of a particle of size i, equation 5.5 can be rewritten into [114, 125]:

$$\beta_{i,j} = \epsilon_{i,j} \left(\frac{3m_1}{4\pi\rho_s} \right)^{1/6} \left(\frac{6k_BT}{\rho_s} \right)^{1/2} \left(\frac{1}{i} + \frac{1}{j} \right)^{1/2} \left(i^{1/3} + j^{1/3} \right)^2 \quad (5.6)$$

Dependent on the coagulation regime different formulations of $\beta_{i,j}$ apply. Whereat, the coagulation regime is determined by the ratio of gas mean free path λ_g and particle size d_p, which is described by the Knudsen number[2]

$$\mathrm{Kn} = 2\lambda_g/d_p \quad (5.7)$$

with

$$\lambda_g = \frac{k_BT}{\sqrt{2}\pi d_g^2 p} \quad (5.8)$$

where d_g denotes the gas molecule diameter [116].

For *particle inception* only one PAH molecule size class needs to be considered and equation 5.4 reduces to its second term [125].

$$\frac{dN_{i=2PAH,PI}}{dt} = \beta_{i_{PAH},j_{PAH}} N_{PAH}^2 \quad (5.9)$$

Due to the small size of PAH molecules, only the free molecular coagulation regime is relevant ($\mathrm{Kn} \to \infty$), where the collision frequency factor is determined by[3]

$$\beta_{i_{PAH},j_{PAH}} = \epsilon_{i,j} \left(\frac{3m_1}{4\pi\rho_s} \right)^{1/6} \left(\frac{6k_BT}{\rho_s} \right)^{1/2} 4\sqrt{2}\, i_{PAH}^{1/6}. \quad (5.10)$$

The Smoluchowski equation for the *PAH condensation* process [125], which occurs on soot particles of all size classes, reads

$$\frac{dN_i}{dt} = \beta_{i_{PAH},i-i_{PAH}} N_{PAH} N_{i-i_{PAH}} - \beta_{i_{PAH},i} N_{PAH} N_i. \quad (5.11)$$

---

[1]Herein, the mass contribution of hydrogen is neglected.

[2]For Kn $\ll$ 1, the continuum regime is present, in which the soot particles slip through the gas phase molecule fluid. For Kn $\to \infty$, the particles act like free molecules and the free molecular coagulation regime is present. The regime in between (Kn $\approx$ 1) is called transition regime.

[3]For detailed derivation see [125].

Assuming $j \ll i$, the collision frequency factor is estimated with

$$\beta^{fm}_{i_{PAH}, j_{PAH}} = \epsilon_{i,j} \left(\frac{3m_1}{4\pi\rho_s} \right)^{1/6} \left(\frac{6k_B T}{\rho_s} \right)^{1/2} i^{2/3} j^{-1/2}. \quad (5.12)$$

The *coagulation* of soot particles with other soot particles is described using the Smoluchowski equation 5.4 without any simplification [125]. Dependent on the soot particle size and surrounding environment, the particles either combine to larger spherical particles or form agglomerates [126]. Hence, the collision frequency factor needs to be modelled for all coagulation regimes [125]. In the free molecular coagulation regime, the collision frequency factor is calculated according to equation 5.12. In the continuum regime, $\beta_{i,j}$ reads

$$\beta^c_{i,j} = \frac{2k_B T}{3\eta} \left(i^{1/3} + j^{1/3} \right)^2 \left(\frac{1 + 1.257\mathrm{Kn}_i}{i^{1/3}} + \frac{1 + 1.257\mathrm{Kn}_j}{j^{1/3}} \right)^{1/2} \quad (5.13)$$

with η as gas viscosity and Kn_i as Knudsen number of particles with size i. The collision frequency factor in the transition coagulation regime is approximated using a harmonic weight of $\beta^{fm}_{i,j}$ and $\beta^c_{i,j}$.

The heterogeneous surface reactions (*surface growth* and *oxidation*) are determined by the chemical reaction rates and concentrations of the gas phase species [116]. The calculation of the surface reactions bases on an active site approach [125], in which the number of active sites is assumed to be proportional to the soot surface [121]. On the active site, Hydrogen-Abstraction, C_2H_2-Addition, and Ring-Closure (HACARC) reactions take place[1], which lead to surface growth of soot particles [125]. Whereas, oxidation reactions by O_2 and OH on the active sites lead to soot mass reduction.

5.4.2 Method of Moments

The modelling of the soot PSDF in form of the individual particle size classes leads to an infinite number of partial differential equations for the number density [138] in the form of

$$\rho \frac{\partial N_i/\rho}{\partial t} - \rho \frac{\chi}{2\mathrm{Le}_i} \frac{\partial^2 N_i/\rho}{\partial Z^2} = \omega_{N_i} \quad (5.14)$$

---

[1]The HACARC mechanism is a modification of the HACA mechanism of Frenklach [51]. A detailed description can be found in [114].

with

$$\omega_{N_i} = \frac{dN_{Particle\,Inception}}{dt} + \frac{dN_{Condensation}}{dt} + \frac{dN_{Coagulation}}{dt} \qquad (5.15)$$
$$+ \frac{dN_{Surface\,growth}}{dt} + \frac{dN_{Oxidation}}{dt}$$

as the convective diffusive operator for particle number density of size class i [113]. The equation system can be reduced to a limited number of equations using the method of moments[1] [52]. In this approach, the mathematical moments

$$M_r = \sum_{i=1}^{\infty} i^r N_i \qquad r = 0, 1, ..., \infty \qquad (5.16)$$

of the PSDF are determined instead of calculating the particle number in each particle size class [125]. In equation 5.16, N_i denotes the number density of soot particles of size i defined with $m_i = i\,m_1$.

The first statistical moments are coupled with physical characteristics of the soot PSDF [125]. The zeroth moment

$$M_0 = \sum_{i=1}^{\infty} N_i = N \qquad (5.17)$$

represents the total number density of particles. The first moment

$$M_1 = \sum_{i=1}^{\infty} i N_i = f_V \frac{\rho_s}{m_1} \qquad (5.18)$$

is related to the mean mass f_m or mean volume fraction f_V of the particles[2]. The second moment

$$\sigma^2 = \frac{M_2}{M_0} - \left(\frac{M_1}{M_0}\right)^2 \qquad (5.19)$$

reflects the variance of the soot PSDF, and the third moment gives information about the skewness of the PSDF. Using the zeroth and first soot moments and assuming spherical particles, the particle mean

---

[1]The method of moments bases on the assumption that any mathematical function can be described by its statistical moments.
[2]The soot density ρ_s is assumed to be equal to 1800 kg/m^3.

diameter

$$d_{Soot} = \sqrt[3]{\frac{6}{\pi \rho_s} \frac{M_{W,C_1}}{N_A} \frac{M_1}{M_0}} \qquad (5.20)$$

can be derived [179], where M_{W,C_1} is the molecular weight of the smallest unit occurring in soot particles, and N_A is the Avogadro constant.

The moments of the soot PSDF can be derived as a function of mixture fraction Z. The non-stationary laminar flamelet equation [121] of the r-th density weighted statistical moment $\overline{M_r} = M_r / \rho N_A$ under the assumption of unity Lewis number reads

$$\rho \frac{\partial \overline{M_r}}{\partial t} - \rho \frac{\chi}{2} \frac{\partial^2 \overline{M_r}}{\partial Z^2} = \omega_r. \qquad (5.21)$$

The source term ω_r accounts for the outlined processes of particle inception, condensation, coagulation, surface growth and oxidation.

Further details about the method applied are found in [52, 125].

5.5 Coupling to 3D-CFD Code

The coupling scheme between the flamelet code and the 3D-CFD code is illustrated in figure 5.2.

The 3D-CFD code solves the energy equation of the favre-averaged total enthalpy $\widetilde{H}$.

The flamelet concept requires, additionally to the gas phase, liquid phase, and combustion progress variable conservation equations, the solution of the transport equations[1] for the favre-averaged mixture fraction $\widetilde{Z}$ and its variance $\widetilde{Z''^2}$ [12].

$$\frac{\partial \left(\overline{\rho} \widetilde{Z} \right)}{\partial t} + \nabla \cdot \left(\overline{\rho} \widetilde{\mathbf{v}} \widetilde{Z} \right) = \nabla \cdot \left(\overline{\rho v'' Z''} \right) + \overline{\rho} \dot{\omega}_{Spray} \qquad (5.22)$$

$$\frac{\partial \left(\overline{\rho} \widetilde{Z''^2} \right)}{\partial t} + \nabla \cdot \left(\overline{\rho} \widetilde{\mathbf{v}} \widetilde{Z''^2} \right) = \nabla \cdot \left(\frac{\mu_t}{\mathrm{Sc}_{\widetilde{Z''^2}}} \cdot \nabla \widetilde{Z''^2} \right) + \frac{2\mu}{\mathrm{Sc}_{\widetilde{Z''^2}}} \left(\nabla \widetilde{Z} \right)^2 - \overline{\rho} \widetilde{\chi} \qquad (5.23)$$

In equation 5.22, the source term $\dot{\omega}_{Spray}$ accounts for fuel mass evaporating from spray. The turbulent transport term $\widetilde{v'' Z''}$ is closed using a

---

[1]A detailed derivation can be found in [65].

gradient flux assumption.

$$\widetilde{v''Z''} = -\frac{\nu_t}{\mathrm{Sc}_t} \cdot \nabla \widetilde{Z} \qquad (5.24)$$

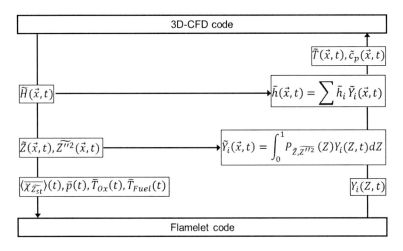

Figure 5.2: Coupling scheme between flamelet code and 3D-CFD solver according to [12]

Furthermore, the 3D-CFD code provides the flamelet parameter, i.e. scalar dissipation rate $\left\langle \widetilde{\chi_{\widetilde{Z}_{st}}} \right\rangle$, mean pressure $\bar{p}$, mean oxidizer temperature $\overline{T}_{ox}$, and mean fuel vapour temperature $\overline{T}_{fuel}$ [12]. In this work, the oxidizer temperature corresponds to the mean unburnt mixture temperature, and the fuel temperature is assumed to a constant value, i.e. $T_{fuel} = 350$ K. Note that the pure fuel temperature has only a minor impact on the flamelet solution in mixture space considered in the following.

The favre-averaged scalar dissipation rate reads

$$\widetilde{\chi} = C_\chi \frac{\widetilde{\varepsilon}}{\widetilde{k}} \widetilde{Z''^2} \qquad C_\chi = 2.0 \qquad (5.25)$$

where C_χ describes the ratio of scalar and velocity dissipation timescales. The scalar dissipation rate is conditioned at stoichiometric mixture fraction and reads in volume average form

$$\langle \overline{\chi_{\widetilde{Z}_{st}}} \rangle = \frac{\int \langle \chi_{\widetilde{Z}_{st}} \rangle \rho \, P_{\widetilde{Z}, \widetilde{Z''^2}, Z_{st}} \, \mathrm{d}V}{\int \rho \, P_{\widetilde{Z}, \widetilde{Z''^2}, Z_{st}} \, \mathrm{d}V}. \tag{5.26}$$

The flamelet equations 5.1 and 5.2 are solved in a separate flamelet code, which is coupled to the 3D-CFD code interactively. The flamelet parameters governing the unsteady development over the flamelet are extracted from the 3D-CFD code by statistical averaging over the representative domain using a presumed β-PDF

$$P_{\widetilde{Z}, \widetilde{Z''^2}} = Z^{\alpha-1}(1-Z)^{\beta-1} \frac{\Gamma(\alpha+\beta)}{\Gamma(\alpha)\,\Gamma(\beta)} \tag{5.27}$$

with

$$\alpha = \widetilde{Z}\left(\frac{\widetilde{Z}(1-\widetilde{Z})}{\widetilde{Z''^2}} - 1\right) \qquad \beta = (1-\widetilde{Z})\left(\frac{\widetilde{Z}(1-\widetilde{Z})}{\widetilde{Z''^2}} - 1\right), \tag{5.28}$$

whose shape is determined by the mean values of $\widetilde{Z}$ and its variance $\widetilde{Z''^2}$.

The calculated instantaneous mass fractions $Y_i(Z,t)$ are transformed from mixture fraction space to physical space, i.e. to Favre averaged mean mass fractions $\widetilde{Y}_i(\mathbf{x}, t)$, [169] according to

$$\widetilde{Y}_i(\mathbf{x}, t) = \int P_{\widetilde{Z}, \widetilde{Z''^2}}(Z)\, Y_i(Z, t)\, \mathrm{d}Z. \tag{5.29}$$

The coupling between the flamelet code and the 3D-CFD code is done by calculating the local temperature $\widetilde{T}$ and the local specific heat capacity $\widetilde{c}_p$ in the flamelet code based on enthalpy of formation $\widetilde{h}_i$ and species mass fractions $\widetilde{Y}_i$. The calculation requires the retrieval of the total enthalpy $\widetilde{H}$ from the 3D-CFD code. The updated temperature $\widetilde{T}$ is then calculated in an iterative procedure according to

$$\widetilde{H}(x, t) = \sum_{i=1}^{N} h_i \, \widetilde{Y}_i + \widetilde{c}_p \, \mathrm{d}\widetilde{T}. \tag{5.30}$$

The calling of the flamelet code by the 3D-CFD code takes place every time step for each grid cell possessing a combustion progress greater than 0.95. The flamelet solution bases thereby on a single flamelet modelling approach. In order to account for different species concentrations in different grid cells, conditional mixing is applied

$$Y_{i,Z,b} = \frac{P_{\widetilde{Z}_{old},\widetilde{Z''}^2_{old}} \, m_{old,b} \, Y_{i,old,b} + P_{\widetilde{Z}_{new},\widetilde{Z''}^2_{new}} \, m_{new,b} \, Y_{i,new,b}}{P_{\widetilde{Z},\widetilde{Z''}^2} \, m_{old,b} + P_{\widetilde{Z},\widetilde{Z''}^2} \, m_{new,b}} \qquad (5.31)$$

where the subscripts *old* and *new* refer to the old and new mass burnt.

In analogy to the treatment of the chemical species, the non-stationary flamelet equations for the soot moments (see equation 5.21) are solved by the flamelet code only.

5.6 Flamelet Initialisation

The flamelet solver is used to model the emission formation in burnt zone regions and called for calculation cells possessing a reaction progress greater than 0.95. This demands for an initialisation of the flamelet as burning or partially burnt. The structure of a burning flamelet possessing a reaction progress of $c = 0.95$ is displayed in figure 5.3.

For the flamelet initialisation in burnt zone region, Dekena [43] suggested a Burke-Schumann start condition[1]. The Burke-Schumann condition represents the structure of a full burnt flamelet in mixture fraction space by considering products of stoichiometric combustion, i.e. N_2, CO_2, and H_2O, surplus fuel and O_2 on the fuel rich and lean sides, respectively. This procedure does not account for fuel and state variable dependent species composition and causes high instability issues of the flamelet solver.

A more accurate approach constitutes a flamelet initialisation based on a look-up table by storing the species composition data for the defined reaction progress in the IPV library. However, following this approach, two problems arise. The first one is a lack of data[2] between upper library entry and $Z = 1.0$. The second problem are instability issues of the flamelet solver arising due to the flamelet initialisation based on 0D homogeneous reactor calculations, which are significantly different from 1D flamelet solutions.

---

[1]The author remarks the incorrectness of this approach, but assumes that a physically correct flamelet solution is reached fast due to fast high temperature chemistry.

[2]In case the IPV library covers the range $0.2 \leq \phi \leq 4.0$, a lack of data arises between $Z_{IPV,max} = 0.21016$ and $Z = 1.0$.

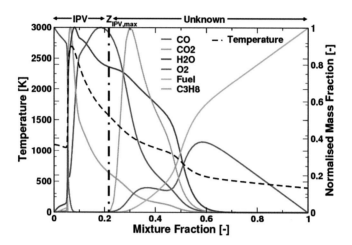

Figure 5.3: Species distribution in mixture fraction space of a burning flamelet with reaction progress of 0.95

The lack of data between $Z_{IPV,max}$ and $Z = 1.0$ can be overcome by extrapolating the library entries in mixture fraction space. From figure 5.3 it can be seen that an extrapolation procedure may generate plausible values for species with their single maximum between $0 \leq Z \leq Z_{IPV,max}$. However, the procedure fails for species with a maximum in the range $Z_{IPV,max} \leq Z \leq 1.0$. The wrong species profile prediction results in instability issues of the flamelet solver, too.

The second problem can be addressed by applying a diffusion step on the species profiles provided by the IPV library. However, the diffusion step makes only sense in case of correct data prediction in the range of $Z_{IPV,max} \leq Z \leq 1.0$.

In order to enable a flamelet initialisation based on IPV library species data, for low scalar dissipation rates the combustion is considered as premixed and the flamelet solver calculates a 0D homogeneous solution with a separated diffusion step in the range $0 \leq Z \leq Z_{max}$, with Z_{max} defined to

$$Z_{max} = \min \left\{ Z_{mean} + 2\, Z''^2, 1 \right\}. \tag{5.32}$$

Thus, for low scalar dissipation rates all data required to initialise the burning flamelet can directly be provided by the IPV library.

The validity of this scalar dissipation rate based approach is investigated in the following. Figures 5.4 and 5.5 compare the temperature and CO mass fraction evolution of the flamelet solution in the range $0 \leq Z \leq Z_{max}$ including separated diffusion step with the full explicit 1D flamelet solution for different scalar dissipation rates.

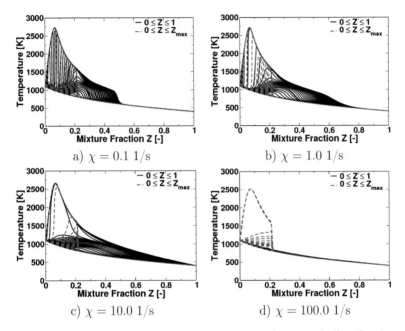

a) $\chi = 0.1 \ 1/\mathrm{s}$ \qquad b) $\chi = 1.0 \ 1/\mathrm{s}$

c) $\chi = 10.0 \ 1/\mathrm{s}$ \qquad d) $\chi = 100.0 \ 1/\mathrm{s}$

Figure 5.4: Comparison of temperature evolution calculated using limited flamelet solution in the range $0 \leq Z \leq Z_{max}$ ($Z_{max} = 0.21016$) with full explicit 1D flamelet solution for different scalar dissipation rates in the time interval $0 \leq t \leq 1$ s

In general, the full explicit flamelet solution shows a smoothing and shifting towards higher Z values of the maximum values of T and CO mass fraction with increasing scalar dissipation rate. This is due to the increasing diffusion impact (see equations 5.1 and 5.2).

For $\chi = 0.1 \ 1/\mathrm{s}$, the temperature and CO mass fraction evolution of the limited flamelet solution matches the full explicit flamelet solution in the considered range $0 \leq Z \leq 0.21016$ well.

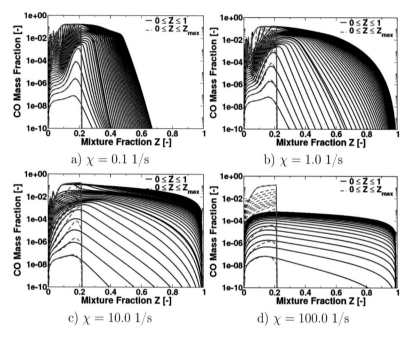

Figure 5.5: Comparison of CO mass fraction evolution calculated using limited flamelet solution in the range $0 \leq Z \leq Z_{max}$ ($Z_{max} = 0.21016$) with full explicit 1D flamelet solution for different scalar dissipation rates in the time interval $0 \leq t \leq 1$ s

At a scalar dissipation rate of 1.0 1/s, small differences in temperature and CO mass fraction evolution occur with increasing time between the limited and the full flamelet solution. However, qualitatively the curves are still comparable.

The differences increase with a scalar dissipation rate of 10.0 1/s. At a scalar dissipation rate of 100.0 1/s the flamelet of the full explicit solution does not ignite due to the high diffusivity induced and the CO mass fraction stays on a low level. In contrast, the 0D homogeneous solution shows an igniting flamelet.

Summarising, the 0D homogeneous solution provides comparable data to the full explicit 1D flamelet solution for scalar dissipation rates smaller than 100 1/s. For scalar dissipation rates greater than 100 1/s, the differences between the 0D homogeneous solution and the full explicit 1D flamelet solution become heavy[1].

5.7 Comparison with Homogeneous Reactor Calculations

In order to examine the basic emission model functionality, the 3D-CFD flamelet solution is compared with 0D homogeneous reactor calculations in the following. Note that the comparison can only base on qualitative arguments. For numerical investigation in 3D-CFD, the simplified test case specified in appendix A.1.2 is used, where the mixture transformation is initiated by auto-ignition. Both, the simplified test case and the homogeneous reactor, are initialised with $\phi = 2.0$, $T = 850$ K, and $\psi = 0.0$. The pressure is varied between $p = 10$ bar and $p = 30$ bar. Figures 5.6 and 5.7 compare the temporal evolution of the NO mass fraction, CO mass fraction, HC mass fraction (where HC is represented by C_3H_8), zeroth statistical soot moment, and first statistical soot moment after auto-ignition, calculated in 0D homogeneous reactor calculation and 3D-CFD flamelet code.

Considering the 0D homogeneous reactor calculations, the NO mass fractions increases continuously over time. Between $p = 10$ bar and $p = 30$ bar small differences of the absolute NO values can be observed, which intensify with increasing time. The CO mass fractions calculated in 0D homogeneous reactor calculations increase until $t = 0.0004$ s. The slope of increase is thereby enhanced with increasing pressure. For values greater than $t = 0.0004$ s, the CO mass fractions of the different pressures converge and stay almost constant. The HC mass fractions evaluated after auto-ignition follow the ranking $Y_{HC,p=30bar} > Y_{HC,p=20bar} > Y_{HC,p=10bar}$. With increasing time, the mass fractions decrease and become almost zero at $t = 0.0003$ s. Whereat, the gradient increases with higher pressure. In 0D homogeneous reactor calculations, the zeroth statistical soot moment, describing the total number density (see equation 5.17), increases strongly shortly after auto-ignition as a result of particle inception. The maximum peak is thereby increased with increasing pressure value. The first statistical soot moment, which describes the mean soot volume fraction (see equation 5.18), peaks slightly after the maximum of the zeroth soot moment. This is due to coagulation processes ongoing, increasing the value of M_1 and decreasing M_0.

---

[1]Note that a mean value of $10 \leq \chi \leq 20$ 1/s is representative for diesel engine combustion.

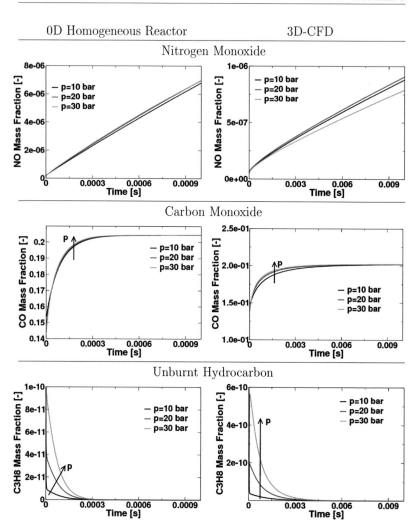

Figure 5.6: Temporal evolution of NO, CO, and HC after auto-ignition calculated in 0D homogeneous reactor calculation and 3D-CFD flamelet code for $\phi = 2.0$, $T = 850$ K, $\psi = 0.0$, and varying p

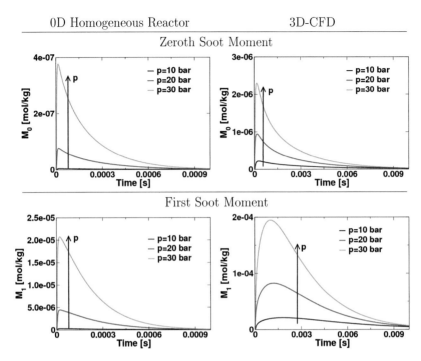

Figure 5.7: Temporal evolution of zeroth and first statistical soot moment after auto-ignition calculated in 0D homogeneous reactor calculation and 3D-CFD flamelet code for $\phi = 2.0$, $T = 850$ K, $\psi = 0.0$, and varying p

After reaching the maximum soot moment value, both M_0 and M_1 start to decrease as a result of soot oxidation dominating the particle inception and coagulation processes. The slope of M_0 is steeper than the one of M_1.

In general, the emissions and soot moments calculated in 3D-CFD flamelet code show a similar behaviour as the one evaluated in 0D homogeneous reactor calculations. However, the absolute values differ. This is due to differences in modelling approach. While the 0D homogeneous reactor calculation bases on constant volume conservation equations, the 3D-CFD flamelet solution bases on constant pressure conservation equations.

Thus, emissions calculated using the interactively coupled flamelet model match qualitatively 0D homogeneous reactor calculations, enabling the fuel-dependent modelling of emission formation in 3D-CFD.

5.8 Premixed Flame Emission Formation in Simplified Test Case

5.8.1 Basic Functionality Study

In the following, the functionality of the emission model in premixed flame combustion is investigated using the simplified test case specified in appendix A.1.2. The simplified test case is homogeneously initialised with $\phi = 2.0$, $T = 750$ K, $p = 20$ bar, and $\psi = 0.0$. Additionally, a wall film with a thickness of $d_{Wf} = 1$ μm is set on the chamber wall. The mixture transformation is initiated by a spark plug, resulting in one flame volume which transforms the in-cylinder charge. However, in premixed SI engine combustion processes, auto-ignition phenomena in front of the propagating flame can lead to the formation of multiple burning volumes (see section 4.6), which needs to be modelled by the single flamelet approach. In order to examine the functionality of the single flamelet modelling approach in separated burning volumes, an additional case is investigated, where the mixture transformation is initiated by two independent spark plugs.

For numerical investigation, the G-equation model is used with the turbulent flame speed formulation D introduced in section 3.7.1 and Perlman's laminar flame speed fitting function.

Figure 5.8 displays the course of the flamelet input parameters, i.e. pressure, oxidizer temperature, mass, and scalar dissipation rate for both cases investigated. In figure 5.9, the calculated emissions and statistical moments are displayed.

Figures 5.10 and 5.11 display the corresponding distributions of the combustion progress variable, temperature, scalar dissipation rate, and soot mass fractions for different times.

In case of premixed combustion initiated by one spark plug, the global in-cylinder pressure (figure 5.8 a) increases continuously with proceeding mixture transformation due to flame propagation. The flame propagation leads to an increase of the unburnt mixture temperature (figure 5.8 b). This temperature constitutes the oxidizer temperature passed to the flamelet code. As the flame propagates through the chamber, the mass contributing to the flamelet increases (figure 5.8 c). The scalar dissipation rate (figure 5.8 d), evaluated in the flamelet domain, possesses relatively high values in the beginning of flamelet calculation at $t = 0.003$ s. This is due to mixture inhomogeneities, which occur close to the chamber wall where the flame is initiated as a result of evaporating wall film. As the flame propagates in direction of the chamber center, the impact of these inhomogeneities on the total scalar dissipation rate decreases. At $t = 0.008$ s a second increase of the scalar dissipation rate can be observed. At this time, the flame reaches the cylinder liner, where inhomogeneities from evaporating wall film occur, leading to an increase of the scalar dissipation rate.

The corresponding zeroth statistical soot moment (figure 5.9 a) increases continuously until reaching a maximum at $t = 0.0075$ s. This time corresponds to the end of rapid burning phase, which becomes apparent from pressure trace.

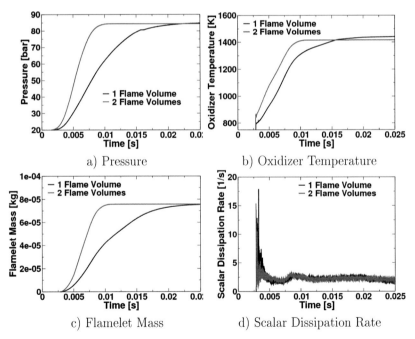

a) Pressure   b) Oxidizer Temperature

c) Flamelet Mass   d) Scalar Dissipation Rate

Figure 5.8: Flamelet model input parameters calculated in simplified test case for one flame volume and two separated flame volumes

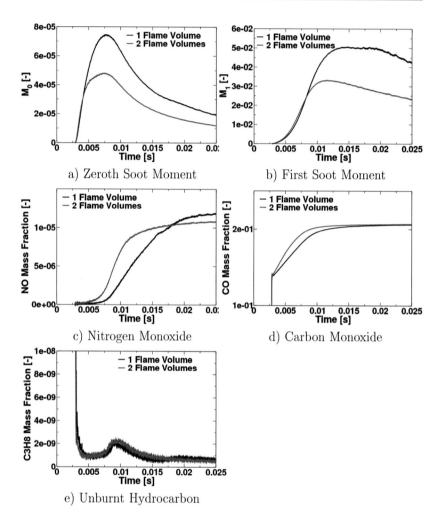

a) Zeroth Soot Moment

b) First Soot Moment

c) Nitrogen Monoxide

d) Carbon Monoxide

e) Unburnt Hydrocarbon

Figure 5.9: Emissions and soot moments calculated in simplified test case for one flame volume and two separated flame volumes

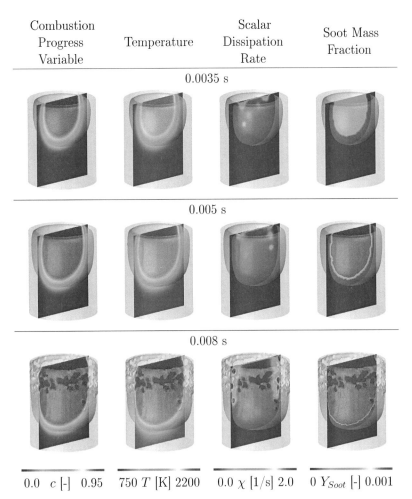

Combustion Progress Variable | Temperature | Scalar Dissipation Rate | Soot Mass Fraction

0.0035 s

0.005 s

0.008 s

0.0 c [-] 0.95   750 T [K] 2200   0.0 χ [1/s] 2.0   0 Y_{Soot} [-] 0.001

Figure 5.10: Distributions of combustion progress variable, temperature, scalar dissipation rate, and soot mass fraction calculated in simplified test case for one flame volume

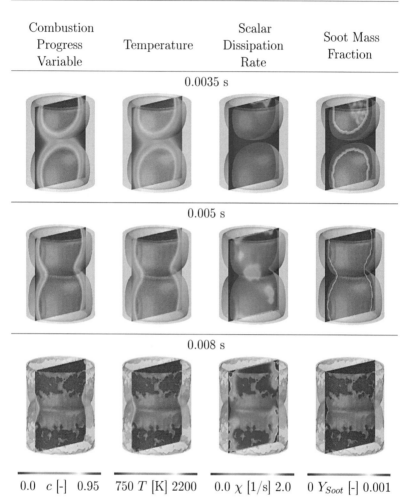

Combustion Progress Variable	Temperature	Scalar Dissipation Rate	Soot Mass Fraction

0.0035 s

0.005 s

0.008 s

0.0 c [-] 0.95   750 T [K] 2200   0.0 χ [1/s] 2.0   0 Y_{Soot} [-] 0.001

Figure 5.11: Distributions of combustion progress variable, temperature, scalar dissipation rate, and soot mass fraction calculated in simplified test case for two separated flame volumes

Proceeding to the maximum of M_0, the moment decreases due to particle inception dominating and continuing oxidation reactions, and coagulation processes. Latter ones lead to a continuous increase of M_1 (figure 5.9 b). Whereat, M_1 peaks slightly after the maximum of the zeroth soot moment. After $t = 0.02$ s, oxidation reactions lead to a decrease of M_1, too. The NO (figure 5.9 c) and CO (figure 5.9 d) mass factions increase continuously, while HC (figure 5.9 e) decreases in the initial flame phase. However, the further trace of HC shows a second peak of the mass fraction at $t = 0.008$ s. At this time, the flame reaches the cylinder liner and the evaporating wall film leads to an increase of fuel-air equivalence ratio and thus HC emission.

In case of two separately initialised flame fronts, the total flame volume and thus transformation rate is enhanced. As a result, the global pressure (figure 5.8 a) as well as the oxidizer temperature (figure 5.8 b) values increase with a greater slope in comparison with mixture transformation by one flame. The enhanced transformation rate is reflected in the mass growth rate in the flamelet domain (figure 5.8 c). Since both cases are initialised homogeneously (apart from the wall film), the different transformation configurations do not have an impact on the scalar dissipation rate (figure 5.8 d). As soon as the flame front is close to the combustion chamber wall, a temporary increase can be observed, which occurs in case of mixture transformation by two flames at an earlier time due to the increased flame propagation speed. Additionally, the slope of scalar dissipation rate decrease is enhanced.

In comparison with mixture transformation by one flame volume, the zeroth and first statistical soot moments evaluated for the case of mixture transformation by two flame volumes are decreased (figures 5.9 a and b). This is due to the increased pressure and temperature values occurring in case of mixture transformation by two flames. Increased pressure and temperature values lead to an increase of both particle inception as well as soot oxidation processes. Moreover, in case of transformation by two flame volumes the maxima of M_0 and M_1 occur at an earlier time due to the increased burning rate. The trace of the NO and CO mass fraction follows directly the one of the temperature. As a result, NO is formed faster in case of mixture transformation by two flame volumes, but reaches lower values at the end of calculation. However, the CO mass fractions at the end of calculation are the same for both types of transformation, indicating that the CO formation is less sensitive towards temperature than the NO formation. The increase of HC mass fraction at $t = 0.008$ as the flame reaches the cylinder liner is greater in case of mixture transformation by two flame volumes. In contrast, in the end of calculations, the HC mass fraction is lower, which can be ascribed to the increased burning rate, too.

Thus, the interactively coupled flamelet model is a tool, which enables the modelling of fuel-dependent turbulent premixed flame emission formation in 3D-CFD under representative boundary conditions.

Moreover, the single flamelet modelling approach successfully works in modelling of separated burning volumes.

5.8.2   Impact of Evaporating Wall Film

As described in section 5.6, the flamelet model applied in this work bases on an operator splitting approach. The results obtained with this approach only match the full 1D flamelet solution for low scalar dissipation rates ($\chi < 100$ 1/s).

The scalar dissipation rate increases with mixture non-homogeneity. A common known source of non-homogeneity in turbocharged direct injection SI engine combustion is the wall film formed by spray impingement. The impact of this on scalar dissipation rate and emission formation is investigated in the following using the simplified test case specified in appendix A.1.2. The boundary conditions correspond to the test case examined in section 5.8.1 possessing one flame volume. For the investigation of wall film thickness impact on χ and emission formation, the thickness is varied between $d_{Wf} = 1$ μm to $d_{Wf} = 4$ μm. Last value corresponds to chamber regions located directly in the spray target [166]. Figure 5.12 displays the corresponding course of the flamelet input parameters. In figure 5.13, the calculated emissions and statistical moments for the varying wall film thicknesses are displayed.

Figures 5.12 a and b reveal, that an increase of wall film thickness results in a decrease of in-cylinder pressure and temperature. This is due to the increasing fuel-air equivalence ratio resulting from increasing fuel evaporation with wall film thickness, which increases the specific heat capacity of the mixture. Additionally, the enthalpy of evaporation leads to a decrease of the in-cylinder temperature and pressure values. The increased fuel-air equivalence ratio values and temperatures lead to a decrease of flame propagation speed (see section 3.6) and thus mass addition rate to the flamelet domain (figure 5.12 c) in the flame propagating phase until $t = 0.008$ s. At $t = 0.008$ s the flame reaches the wall and the evaporating wall film contributes continuously to the flamelet mass, which leads to an increase of flamelet mass with increasing wall film thickness. The increased wall film evaporation rate with increasing wall film thickness is also reflected in the values of the scalar dissipation rate (figure 5.12 d). As the flame reaches the chamber wall, the increasing mixture non-homogeneities with wall film thickness lead to an increase of χ. The maximum values are thereby $\chi = 3.5$ 1/s for $d_{Wf} = 1$ μm, $\chi = 6.5$ 1/s for $d_{Wf} = 2$ μm, $\chi = 27.5$ 1/s for $d_{Wf} = 3$ μm, and $\chi \geq 100$ 1/s for $d_{Wf} = 4$ μm.

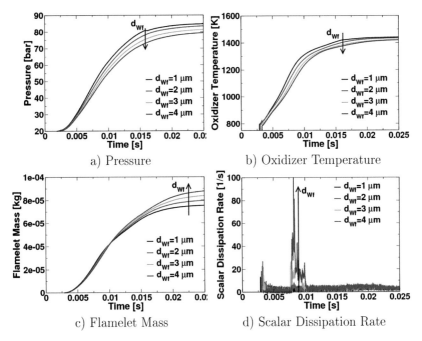

Figure 5.12: Flamelet model input parameters calculated in simplified test case for varying wall film thicknesses

As a result of decreased temperature and pressure, and increased fuel-air equivalence ratio with increasing wall film thickness, the zeroth statistical soot moment increases (figure 5.13 a). Moreover, the moment peaks at a later time with increasing wall film thickness due to the continuing increased mass add to the flamelet domain. Last effect leads to a continuing increase of M_1 for $d_{Wf} = 3$ μm and $d_{Wf} = 4$ μm (figure 5.13 b), while $d_{Wf} = 1$ μm possesses a maximum. Due to the increased mass add with increasing wall film thickness, soot formation processes dominate increasingly oxidation processes. The NO (figure 5.13 c) mass fraction decreases with increasing d_{Wf}. The NO formation is directly linked to the flame temperature. Thus, with decreasing flame temperature as a result of increasing fuel-air equivalence ratio with d_{Wf}, the NO formation is reduced. In contrast, the CO and HC mass fractions increase with d_{Wf} (figures 5.13 d and e) due to increased fuel-air equivalence ratio with d_{Wf}.

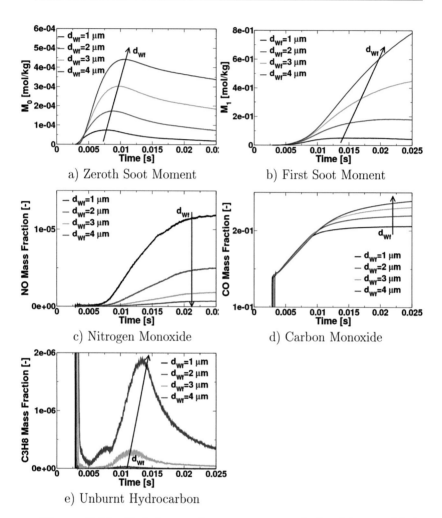

a) Zeroth Soot Moment

b) First Soot Moment

c) Nitrogen Monoxide

d) Carbon Monoxide

e) Unburnt Hydrocarbon

Figure 5.13: Emissions and soot moments calculated in simplified test case for varying wall film thicknesses

Summarising, the model copes effects of mixture non-homogeneities. An increase of fuel-air equivalence ratio due to increasing wall film thickness results thereby in an increase of soot moments, HC, and CO emissions, and a decrease of NO. Increasing the wall film thickness to $d_{Wf} = 3$ μm, the scalar dissipation rate increases to $\chi = 27.5$ $1/\mathrm{s}$.

This value is lower than the critical value $\chi = 100$ 1/s, constituting the boundary between premixed and diffusion flame combustion. However, a wall film thickness of $d_{Wf} = 4$ μm results in an exceed of $\chi = 100$ 1/s. Thus, for wall film thicknesses greater than $d_{Wf} = 3$ μm, the premixed flame assumption is not valid, and the full explicit 1D flamelet needs to be solved. For $d_{Wf} \leq 3$ μm, the operator splitting procedure can be applied. Note that in this investigation the wall film thickness has been initialised on the entire liner. In measurements [166], thicknesses of $d_{Wf} \geq 1$ μm where only measured in regions located inside the spray target.

5.9   Model Validation using Optical Measurements

In the following, the functionality of the coupled G-equation / integrated flamelet / interactive flamelet modelling approach is investigated in terms of emission formation prediction in an optical SI engine. Soot formation processes are promoted by relatively high temperature levels under fuel rich conditions. However, high temperature levels also enhance the soot oxidation process. For this reason, for model validation a part load operating point at 2200 rpm and 0.73 bar manifold pressure is chosen, possessing temperature levels which are high enough to promote the soot formation process, and low enough to slow down soot oxidation process. In test bed measurements, local fuel rich conditions are provoked by an artificial shifting of SOI, i.e. from 250 °CA bTDC to 330 °CA bTDC. Specifications about the optical engine and the operating point can be found in appendix A.6.1 and A.6.2.

For numerical investigation, the G-equation model is used with the turbulent flame speed formulation D introduced in section 3.7.1 and the laminar flame speed fitting function of Perlman.

Figure 5.14 displays the liquid fuel injection and resulting wall film for the two different SOI.

In case of SOI at 250 °CA bTDC, the mean free path of the liquid spray is long due to piston position close to Bottom Dead Center (BDC). As a result, the spray does not directly impinge on the piston and only a small amount of wall film is formed. Additionally, a minor impingement and thus wall film formation can be observed on the liner. By contrast with the SOI at 250 °CA bTDC, the mean free path of the liquid spray of the SOI at 330 °CA bTDC is short. The spray impinges on the piston resulting in a large amount of wall film formed.

SOI 250 °CA bTDC	SOI 330 °CA bTDC

Liquid Fuel

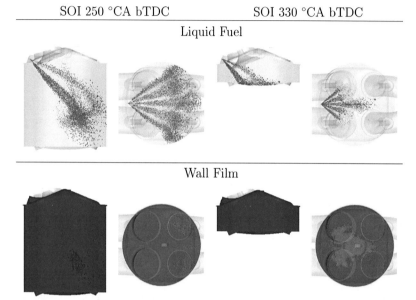

Wall Film

Figure 5.14: Calculated liquid fuel injection (in side and top view at 225 °CA bTDC and 315 °CA bTDC, respectively) and wall film formation (in side and top view at 215 °CA bTDC and 305 °CA bTDC, respectively) for a SOI variation of the optical SI engine at part load operating point (2200 rpm, 0.73 bar)

The difference in wall film formation of the SOI variation are also shown in figure 5.15, which displays the temporal evolution of the total wall film mass.

While the SOI at 250 °CA bTDC leads to the formation of 0.155 mg wall film, the SOI at 330 °CA bTDC forms 3.11 mg. However, the wall film mass decreases continuously due to evaporation and close to spark timing, at 25 °CA bTDC, only a fraction of the initially formed wall film mass remains.

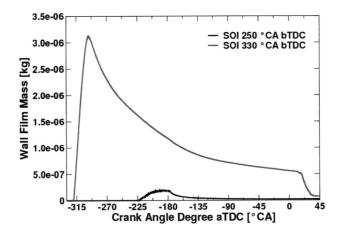

Figure 5.15: Calculated temporal evolution of wall film mass for a SOI variation of the optical SI engine at part load operating point (2200 rpm, 0.73 bar)

The distribution of the remaining wall film for the two different SOI at spark timing is displayed in figures 5.16 and 5.17. Additionally, the figures depict the in-cylinder distribution of the fuel-air equivalence ratio and the temperature.

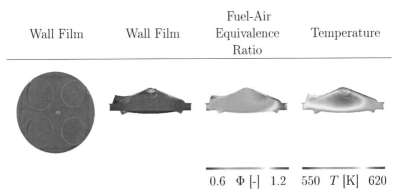

Figure 5.16: Calculated distribution of wall film, fuel-air equivalence ratio, and temperature at 25 °CA bTDC for a SOI at 250 °CA bTDC of the optical SI engine at part load operating point (2200 rpm, 0.73 bar)

Wall Film	Wall Film	Fuel-Air Equivalence Ratio	Temperature

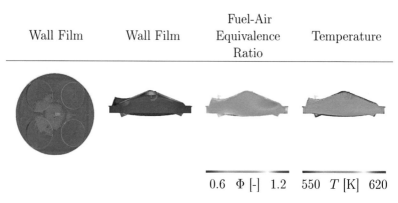

$$0.6 \quad \Phi \ [\text{-}] \quad 1.2 \qquad 550 \quad T \ [\text{K}] \quad 620$$

Figure 5.17: Calculated distribution of wall film, fuel-air equivalence ratio, and temperature at 25 °CA bTDC for a SOI at 330 °CA bTDC of the optical SI engine at part load operating point (2200 rpm, 0.73 bar)

In case of SOI at 250 °CA bTDC (figure 5.16), the wall film is at 25 °CA bTDC located in the bowl close to the liner. The spatial change of the wall film location from the liner into the bowl is a result of piston motion. The in-cylinder distribution of fuel-air equivalence ratio spreads over the range $0.6 \leq \phi \leq 1.2$. Whereat, fuel lean areas are located close to the liner on the intake valve side and close to the exhaust valves. Fuel rich conditions appear close to the liner on the exhaust valve side. The in-cylinder temperature distribution possesses highest values in the cylinder center and decreasing values in direction of the intake valves. In the fuel rich region the temperature level is quite low.

For the SOI at 330 °CA bTDC (figure 5.17), the wall film location at 25 °CA bTDC corresponds with the one shortly after injection. The temperature distribution is uniform. The fuel-air equivalence ratio distribution possesses a fuel lean area close to the exhaust valves. Fuel rich conditions can be found close to the bowl, where the wall film is located. Nevertheless, compared with the SOI at 250 °CA bTDC, the in-cylinder mixture appears to be homogeneously distributed, which can be ascribed to the longer time available for mixture formation process. However, the total area covered with wall film exceeds the one of the SOI at 250 °CA bTDC. The impact of this on combustion process and emission formation is investigated in the following.

Before that, one important issue needs to be mentioned. Usually, in engines the cylinder liner is sealed with the piston using piston rings. However, for optical access, the cylinder liner of the optical engine is made of quartz glass. The piston rings would lead here to a mechanical damage of the glass, which reduces the optical access. For this reason, the optical engine is not equipped with piston rings, but the liner is sealed with a teflon-bronze ring below the glass, which leads to an increase of the fire land and thus the compression volume. Compared with the complete engine, consisting of aluminium cylinders and equipped with piston rings, the resulting compression ratio is much smaller[1]. By modelling the optical engine, the increase of the compression volume is not taken into account[2] and the compression ratio used for calculation agrees with the one of the complete engine. Due to this, the calculated in-cylinder pressure values are higher than the one measured. However, the characteristics are qualitatively reproduced.

Figure 5.18 displays the calculated pressure curves (3D) together with the measured in-cylinder pressure. Furthermore, the pressure curves resulting from thermodynamic analysis (1D) are shown. Last one provides the initial and boundary conditions for the 3D-CFD simulation.

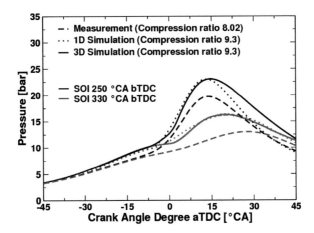

Figure 5.18: Calculated and measured pressure curves for a SOI variation of the optical SI engine at part load operating point (2200 rpm, 0.73 bar)

---

[1]The compression ratio of the optical engine is 8.02, the one of the complete engine is 9.3.

[2]One possibility to account for the increase of compression volume is to adjust the piston position. It is shown in appendix A.6.3, that this kind of adjustment results in a correct prediction of soot formation time, but wrong soot formation location.

The figure reveals that the combustion process of the SOI at 250 °CA bTDC starts considerably before the one of the SOI at 330 °CA bTDC. As a result, the maximal in-cylinder pressure exceeds the one of the SOI at 330 °CA bTDC.

In figure 5.19 the course of the scalar dissipation rate, soot moments, and emissions for the two SOI investigated are displayed. Figures 5.20, 5.21, 5.22, 5.23, 5.24, and 5.25 show the corresponding distributions of the in-cylinder characteristics and emissions.

For the SOI at 250 °CA bTDC (figures 5.20, 5.21, and 5.22), the in-cylinder distribution of the combustion progress variable possesses an initially fast mixture transformation process. However, due to cylinder expansion the mixture transformation is slowed down after 20 °CA aTDC. As a result, the flame does not reach the cylinder wall and unburnt regions remain at 40 °CA aTDC. In the burnt zone region, at 10 °CA aTDC the temperature has high values, which decrease after 20 °CA aTDC due to cylinder expansion. Additionally, heat losses lead to a reduction of the temperature in near-wall regions. However, a relatively high temperature field can be observed close to the liner on the exhaust valve side. In this area, the fuel-air equivalence ratio distribution possesses a marginal fuel rich region, which stays unchanged throughout the whole combustion process.

In line with that, the wall film area does not decrease throughout the combustion. Last finding becomes also apparent from the scalar dissipation rate (figure 5.19 a), which peaks only marginally at 5 °CA aTDC to 0.5 1/s. The resulting soot moments (figure 5.19 b and c) stay close to zero. Only a scaling-up possesses a marginal increase of soot mass starting at 30 °CA aTDC in the area close to the liner on the exhaust valve side, where the mixture is slightly fuel rich and relatively high temperatures predominate.

In contrast, the NO mass fraction (figure 5.19 d) increases continuously and concentrates in high temperature regions nearby the soot formation area. The CO mass fraction (figure 5.19 e) peaks at 5 °CA aTDC and decreases slightly in the further course. Whereat, highest values can be found in the fuel rich area close to the liner on the exhaust valve side. The HC emissions (figure 5.19 f) stay close to zero.

In contrast with the SOI at 250 °CA bTDC, the SOI at 330 °CA bTDC (figures 5.23, 5.24, and 5.25) shows an initially slow mixture transformation process at TDC and 10 °CA aTDC. The mixture transformation is further slowed down due to cylinder expansion and large areas of the cylinder remain unburnt at 40 °CA aTDC. In the transformed area the temperature increases considerably at 10 °CA aTDC, and starts to decrease locally after 20 °CA aTDC in near-wall areas. One reason for this are heat losses through the wall.

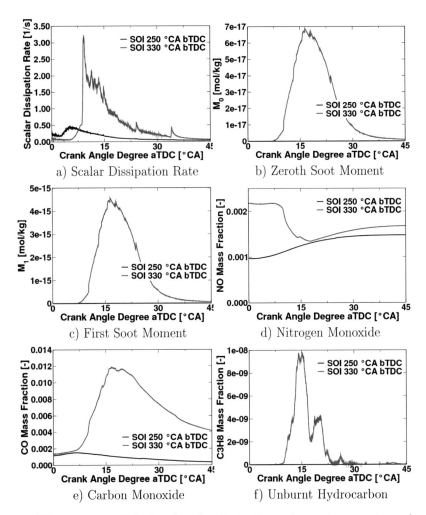

Figure 5.19: Calculated scalar dissipation rate, soot moments, and emissions for a SOI variation of the optical SI engine at part load operating point (2200 rpm, 0.73 bar)

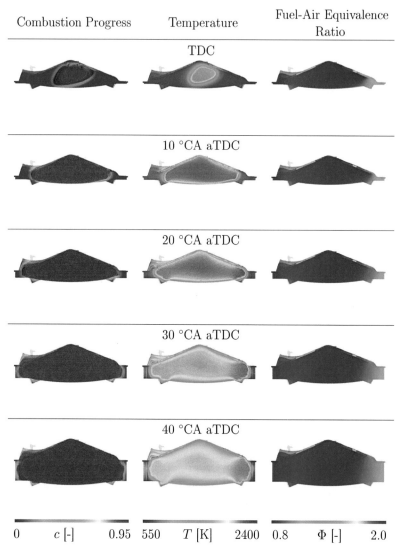

Figure 5.20: Calculated in-cylinder distribution of combustion progress, temperature, and fuel-air equivalence ratio for a SOI at 250 °CA bTDC of the optical SI engine at part load operating point (2200 rpm, 0.73 bar)

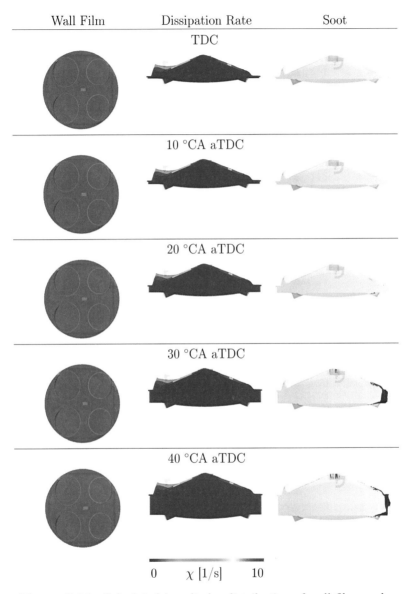

Figure 5.21: Calculated in-cylinder distribution of wall film, scalar dissipation rate, and soot for a SOI at 250 °CA bTDC of the optical SI engine at part load operating point (2200 rpm, 0.73 bar). Soot iso-surface displayed corresponds to a soot mass of 5E-19 kg.

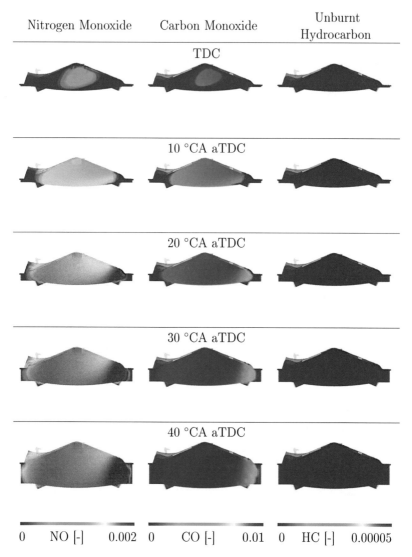

| Nitrogen Monoxide | Carbon Monoxide | Unburnt Hydrocarbon |

0   NO [-]   0.002   0   CO [-]   0.01   0   HC [-]   0.00005

Figure 5.22: Calculated in-cylinder distribution of NO mass fraction, CO mass fraction, and HC mass fraction for a SOI at 250 °CA bTDC of the optical SI engine at part load operating point (2200 rpm, 0.73 bar)

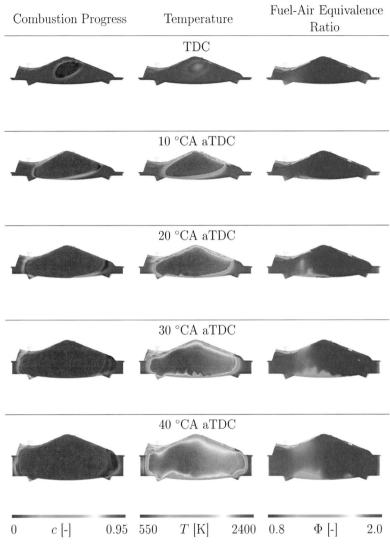

Figure 5.23: Calculated in-cylinder distribution of combustion progress, temperature, and fuel-air equivalence ratio for a SOI at 330 °CA bTDC of the optical SI engine at part load operating point (2200 rpm, 0.73 bar)

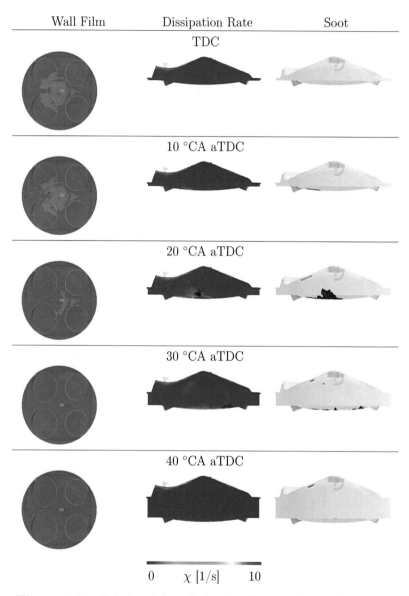

Figure 5.24: Calculated in-cylinder distribution of wall film, scalar dissipation rate, and soot for a SOI at 330 °CA bTDC of the optical SI engine at part load operating point (2200 rpm, 0.73 bar). Soot iso-surface displayed corresponds to a soot mass of 1E-14 kg.

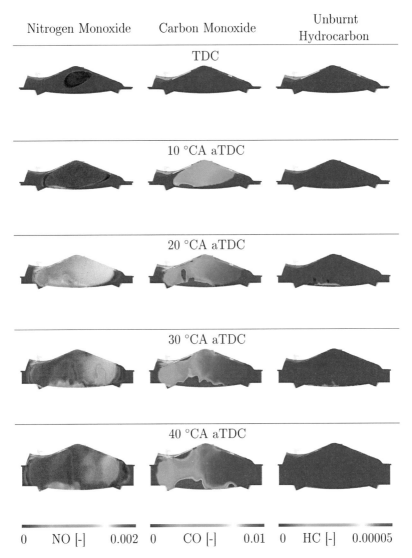

Figure 5.25: Calculated in-cylinder distribution of NO mass fraction, CO mass fraction, and HC mass fraction for a SOI at 330 °CA bTDC of the optical SI engine at part load operating point (2200 rpm, 0.73 bar)

However, the temperature decrease is for the SOI at 330 °CA bTDC not restricted to near-wall areas. It comprehends rather large regions in the combustion chamber starting from the bowl. Nevertheless, compared with the SOI at 250 °CA bTDC, the overall temperature level in the expansion stroke is high and highest temperatures can be found on the intake valve side. The reason for the temperature decrease covering large areas in the cylinder can be found in the wall film distribution. As the flame reaches the bowl in the period 10 °CA aTDC until 30 °CA aTDC, the wall film promptly starts to evaporate, which leads to a local increase of fuel-air equivalence ratio and scalar dissipation rate. Thereby, the fuel-air equivalence ratio reaches locally values greater than 2. In the fuel rich areas the temperature decreases due to an increase of the specific heat capacity as well as an increase of the enthalpy of evaporation. At 40 °CA aTDC, almost the entire wall film is evaporated and the scalar dissipation rate turns zero (figure 5.19 a). In the fuel rich regions resulting from evaporating wall film, soot is immediately formed. Starting from the bowl on the intake valve side at 10 °CA aTDC, the soot expands into the center of the chamber at 20 °CA aTDC. At the same time, the soot moments (figures 5.19 b and c) peak. However, due to ongoing oxidation processes the soot moments decrease and at 30 °CA aTDC, only small amounts of soot can be found in the cylinder bowl. The NO mass fraction (figure 5.19 d) possesses high values at TDC and 10 °CA aTDC in the entire burnt zone. However, at 20 °CA aTDC the overall NO mass faction decreases promptly. The lowest values concentrate in the fuel rich region, where the soot is formed and the temperature level quite low. At 30 °CA aTDC, the NO mass starts overall to increase again, except of the fuel rich regions. Highest values can be found on the intake valve side, where high temperatures predominate. As a result of the higher temperature level in the expansion stroke, the total NO mass formed with a SOI at 330 °CA bTDC exceeds the one of the SOI at 250 °CA bTDC. CO (figure 5.19 e) is formed starting from 10 °CA aTDC primarily in the fuel rich regions. Additionally, a smaller amount of CO can be found on the intake valve side where the highest temperature level predominates. The total CO mass peaks as the soot between 10 °CA aTDC and 20 °CA aTDC and decreases continuously in the further course. The amount of CO formed considerably exceeds the one of the SOI at 250 °CA bTDC. The same is true for HC (figure 5.19 f). However, in contrast to CO, the area where HC is formed is restricted to the fuel rich conditions and low temperatures in the cylinder bowl.

In the following, the results obtained are compared with optical measurements carried out using high-speed-video equipment as well as exhaust gas particle counter measurements. Since for optical measurement no additional lightening is used, the illumination of the pictures is a result of chemiluminescence of molecules from premixed flame and luminescence of soot particles from diffusion flame [14]. The results of the optical measurements are displayed in figure 5.26.

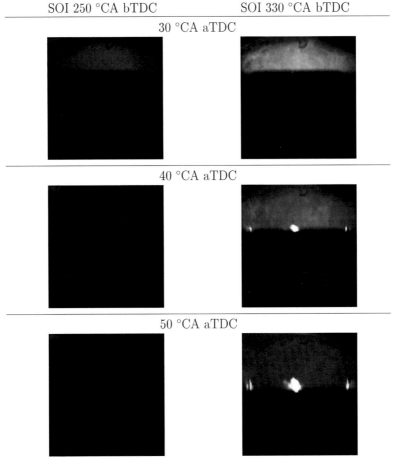

Figure 5.26: Optical measurements for a SOI variation of the optical SI engine at part load operating point (2200 rpm, 0.73 bar)

The figure shows for a SOI at 250 °CA bTDC a marginal luminescence resulting from the premixed flame at 30 °CA aTDC. Afterwards, the premixed flame chemiluminescence exceeds the visible spectrum. By comparison with the SOI at 250 °CA bTDC, the premixed flame chemiluminescence is much more intense at 30 °CA aTDC for a SOI at 330 °CA bTDC. The premixed flame chemiluminescence decreases continuously until 50 °CA aTDC. The longer lasting chemiluminescence of the SOI at 330 °CA bTDC is due to the retarded combustion process. Additionally, soot particle luminescence becomes visible in the center of the bowl starting at 40 °CA aTDC and intensifies at 50 °CA aTDC[1].

The results of the particle counter measurements are shown in table 5.1.

SOI [°CA bTDC]	Particle Number [$1/cm^3$]
250	153000
330	662000

Table 5.1: Measured particle numbers for a SOI variation of the optical SI engine at part load operating point (2200 rpm, 0.73 bar)

In contrast to the optical measurement, the 3D-CFD calculation predicts a marginal soot amount for the SOI at 250 °CA bTDC, which is formed close to the liner on the exhaust valve side. Note that the value of soot mass displayed for the SOI at 250 °CA bTDC (figure 5.21) is much smaller than the value displayed for the SOI at 330 °CA bTDC (figure 5.24). However, the particle counter measurements confirm the formation of a relatively small amount of soot particles for the SOI at 250 °CA bTDC. For the SOI at 330 °CA bTDC, both, the optical measurement and the 3D-CFD calculation show a soot formation located in the cylinder bowl and growing up into the cylinder center[2]. The calculated soot mass for the SOI at 330 °CA bTDC exceeds the soot mass calculated for the SOI at 250 °CA bTDC considerably. This tendency is confirmed by the particle counter measurements, although the relative difference of the particle numbers measured is in comparison with the difference of calculated numbers smaller.

---

[1]Note that the luminescence on the intake and exhaust valve side is a result of light reflection on the glass cylinder.

[2]The time of soot formation calculated does not match the soot formation time resulting from optical measurements. This can be assigned to the difference in compression ratio between measurement and calculation. The impact of compression ratio on soot formation time is investigated in detail in appendix A.6.3.

Summarising, the calculations show that a SOI at 250 °CA bTDC leads to the formation of a marginal wall film amount and a non-homogeneous mixture distribution at TDC. As a result, a relatively high amount of NO, and a small amount of CO and soot are formed. Thereby, the emissions formed are located in a marginal fuel rich region close to the exhaust valves. In contrast, a SOI at 330 °CA bTDC results in a relatively high amount of wall film and a homogeneous mixture distribution at TDC. During the combustion process, the remaining wall film evaporates into the premixed flame, which leads to the formation of a large amount of CO, HC, and soot. Additionally, a relatively large amount of NO is formed in high-temperature regions. The soot emissions concentrate in near-wall regions in the bowl, which is confirmed by the optical measurements. Furthermore, the soot emissions formed exceed the one of the SOI at 250 °CA bTDC considerably. Last finding is in-line with the results obtained from particle counter measurements, although the relative difference of the particle numbers measured is in comparison with the difference of the calculated particle numbers smaller. For a more accurate prediction of the particle number difference, a multiple flamelet approach is required, which solves the flamelet equations rather dependent on local boundary conditions than on averaged in-cylinder values. Additionally, the optical engine measurements base on a standard gasoline fuel, containing a multitude of low and high volatile fuel components. In contrast, the calculations base on an *iso*-Octane reaction mechanism. The accuracy of prediction can be enhanced, using a reaction mechanism which covers the spread of low and high volatile fuel components. Thus, the coupled *G*-equation / integrated flamelet / interactive flamelet modelling approach enables the qualitative prediction of emission formation in direct injection SI engines.

Chapter 6

Summary and Conclusions

The three-dimensional Computational Fluid Dynamic (3D-CFD) development of future Spark Ignition (SI) engines demands for the integration of detailed chemistry, enabling the prediction of fuel-dependent SI engine combustion in all of its complexity. This work shows the potential of a combined G-equation / integrated flamelet / interactive flamelet approach to model the variety of SI engine combustion phenomena, by incorporating detailed chemistry and possessing low computational costs.

The underlying assumption of the combined G-equation / integrated flamelet / interactive flamelet modelling approach is that the inner flame layer is not perturbed by the turbulence, which enables the decoupling of chemistry and turbulent flow modelling. The validity of this assumption is proven for turbocharged direct injection SI engines using the Borghi-Peters diagram. It is shown, that independently of engine examined, the turbulent premixed combustion takes place in the thin reaction zone and corrugated flamelets regimes, where the flamelet assumption holds. Furthermore it is shown, that the flamelet assumption holds for the entire combustion process of turbocharged SI engines. Both, the burning zone and the post flame burning takes place in the thin reaction zone and corrugated flamelets regimes. In the wrinkled flamelet regime, the flame extinguishes.

The G-equation approach is used for flame propagation modelling. The driving force of this approach is the turbulent flame speed, which is primarily defined by the turbulence, but also by the fuel-dependent laminar flame speed. For laminar flame speed closure, different fitting functions are examined and compared with direct numerical calculations. It is shown, that the fitting function of Metghalchi and Keck as well as the formulation of Perlman predict the laminar flame speed in the range $0.8 \leq \phi \leq 1.2$ best. The formulation of Gülder as well as the formulation of Mallard and LeChatelier fail to predict the laminar flame speed adequate enough. On the fuel lean and fuel rich side, the formulation of Perlman shows the smallest error in representing the direct numerically calculated laminar flame speed. Using the simplified test case it is further shown, that under turbulent conditions the maximal in-cylinder pressure differs heavily using the different laminar flame speed fitting functions, especially for fuel rich conditions. The differences in maximal in-cylinder pressure prediction are also found in engine test case, although reduced. However, with decreasing turbulence the impact of the laminar flame speed on turbulent flame propagation increases. As the turbulence decreases heavily in the expansion stroke where the flame is close to the cylinder wall, the influence of the laminar flame speed becomes primarily apparent in terms of near-wall mixture transformation. Here, the greatest difference for the laminar flame speed formulations occurs under fuel rich conditions. Latter investigation emphasises the necessity of correct laminar flame speed modelling on the fuel rich and fuel lean side. Also for turbulent flame speed closure different formulations are examined and compared with measurement data. In the group of turbulent flame speed closure formulations which base on velocity scales the formulation proposed by Abraham et al. and Kawanabe et al. best matches the measurement data in the entire combustion regime. For the turbulent flame speed closure formulations which base on velocity and length scales, the formulations of Kolla et al. and Peters best match the measurement data. Whereat, with both formulations, best reproduction of the measured flame speed is obtained in the wrinkled flamelets regime. In the corrugated flamelets and thin reaction zone regime, Peters' formulation slightly underestimates the measurement data, while Kolla's formulation marginally overestimates the measured flame speeds. These findings are emphasised in the engine test case, where Peters' formulation underpredicts the in-cylinder pressure and Kolla's formulation overpredicts the pressure. However, this study is carried out using the laminar flame speed fitting function of Gülder. For a better reproduction of the measured pressure curves a more accurate laminar flame speed fitting function is required. Furthermore, the G-equation model is validated in engine test case for a fuel-air equivalence ratio variation using optical measurements. The calculated flames match the optical measurements in terms of flame shape and local propagation well. Thus, the model enables the calculation of turbulent premixed flame propagation incorporating turbulent and chemical effects.

The modelling of auto-ignition phenomena is carried out using an integrated flamelet approach, the Ignition Progress Variable (IPV). The integrated flamelet approach bases on the tabulation of fuel-dependent reaction kinetics. By introducing a progress variable for the auto-ignition, detailed chemistry is integrated in 3D-CFD. The progress variable can either base on state variables or combustion species. Compared with combustion species based progress variables it is shown, that the integrated latent enthalpy reproduces the auto-ignition process best. Especially for low progress variable values, the combustion species based progress variables do not reproduce the auto-ignition process adequate enough. In order to decrease the computational demand for library creation, the required library extension is examined by analysing the reaction progress in dependence on the tabulated independent variables. It is shown, that the library can be limited in fuel-air equivalence ratio space to $\phi \leq 3.4$, and in temperature space to $T \leq 400$ K. In the range examined ($0.2 \leq \phi \leq 3.8$, $400 \leq T \leq 1200$ K, $5 \leq p \leq 95$ bar, $0.0 \leq \psi \leq 0.9$), on the fuel lean side, for high temperatures, for pressure, and for Exhaust Gas Recirculation (EGR) rate a limitation is not feasible. In general, the results obtained applying tabulation methods strongly depend on the accuracy of tabulation. For this reason, the required independent variable nodes are examined. For the investigation of the combustion progress variable step size, the calculations are compared with 0D homogeneous reactor calculations. The results show that the difference between directly calculated ignition delay time and the data obtained with the integrated flamelet model decreases with increasing progress variable mesh refinement. Furthermore it is shown, that the mesh refinement can be limited to combustion progress variable smaller than 0.05, indicating the great impact of low combustion progress variable values on auto-ignition. The mixture defined independent variable nodes are examined in engine test case and by direct linear interpolation of the tabulated data. It is shown, that the tabulation step size of fuel-air equivalence ratio has a great impact on auto-ignition time calculated due to interpolation errors occurring applying linear interpolation. In order to avoid interpolation errors, a minimal step size of $\Delta \phi_{Tab} = 0.1$ is required. A higher order interpolation procedure enables the tabulation of ϕ in a courser mesh, but the computational demand in 3D-CFD calculation increases thereby heavily. For the mesh refinement investigated, the tabulated step size of the mixture defined independent variables temperature, pressure, and EGR do not have an impact on auto-ignition time calculated. Here, the mesh size can be decreased to $\Delta T_{Tab} = 100$ K, $\Delta p_{Tab} = 20$ bar, and $\Delta \psi_{Tab} = 0.3$. Next to combustion progress, the library provides the change of thermodynamic data involved. The thermodynamic data retrieval can either base on the tabulation of molecular weight and NASA-polynomial coefficients, or the specific heat capacity and temperature. It is shown that the tabulation of c_p and T possesses less interpolation errors than M_W and A_i. In order to model engine knock and Homogeneous Charge Compression Ignition (HCCI) combustion, the integrated flamelet model is coupled with the G-equation model. The functionality of the coupled G-equation / integrated flamelet model is shown in terms of an HCCI combustion. The investigation shows, that the method allows for examining the interaction between flame propagation and locally appearing auto-ignition processes in the unburnt mixture. However, for the prediction of an average HCCI engine combustion process for operating points with high cyclic variations in test bed measurements, multi-cycle calculations are required. Furthermore, the model capability to predict pre-ignition phenomena is investigated in a highly turbocharged SI engine, which is feasible especially for the investigation of this stochastic auto-ignition phenomena. For this investigation, multi-cycle calculations are carried out. The calculated pre-ignition tendency and location matches the measurement data well, although the calculated total number of pre-ignition events in a given time period exceeds the measurement data. This can be ascribed to the wall boundary temperatures which are set to a constant value in the simulation. For a more accurate prediction of the number of pre-ignition events a coupling to temperature field calculations of the solid domain in so called conjugate heat transfer calculations is required. Nevertheless, the investigation shows that pre-ignition phenomena are a result of mixture in-homogeneity.

The modelling of emission formation bases on an interactively coupled flamelet approach, the Transient Interactive Flamelet (TIF) model, which is coupled to the G-equation / integrated flamelet model. The TIF model is used to model the emission formation in burning and burnt zone regions. This demands for a flamelet initialisation as partially burnt. By considering the combustion as premixed and thus the scalar dissipation rate as low, the flamelet model bases on an operator splitting procedure, enabling the flamelet initialisation by tabulated 0D homogeneous reactor solutions. It is shown that this procedure provides comparable results to the full 1D flamelet solution for scalar dissipation rates smaller than 100. Furthermore, a comparison of the emissions calculated using the TIF model with 0D homogeneous reactor calculations shows a qualitatively good agreement. A functionality investigation in the simplified test case shows the potential of the combined G-equation / integrated flamelet / interactive flamelet approach to model the emission formation in premixed flame combustion, even for multiple burning volumes. It is shown that the model also accounts for the impact of evaporating wall film on emission formation. Furthermore, the functionality of the interactive flamelet model to predict emission formation is shown for a variation of Start Of Injection (SOI) possessing different wall film formation tendencies in an optical SI engine. The model predicts higher soot emissions for an earlier SOI, resulting from wall film evaporation into the premixed flame during the combustion process. These results are confirmed by optical measurements, although the relative difference of the particle numbers measured is, in comparison with the difference of the calculated particle numbers, smaller. For a more accurate prediction of the emission formation, a multiple flamelet approach is required, which solves the flamelet equations rather dependent on local boundary conditions than on averaged in-cylinder values. Furthermore, the emission formation in SI engines is strongly dependent on the fuel evaporation rate, which is determined by the hundreds of components a fuel consists of. In this work the calculations base on a single component fuel approach, for both, the liquid and the gaseous phase. For a better emission formation prediction, a multi-component fuel approach is required, which accounts for both, liquid fuel demixing effects and fuel-dependent emission formation in the gaseous phase.

Summarising, the combined G-equation / integrated flamelet / interactive flamelet modelling approach provides an appealing solution to model the variety of SI engine combustion phenomena by incorporating detailed chemistry.

Appendix A

Appendix

A.1 Simplified Test Case

The simplified test case represents an adiabatic pressure vessel without moving mesh. The geometry of the simplified test case is similar to an in-cylinder combustion chamber at BDC, by possessing a length of 0.0474 m and a diameter of 0.0392 m. The grid cells have a length, height, and depth of 0.001 m. The resulting grid number is approximately 114000.

A.1.1 Mesh Size Sensitivity

The mesh size sensitivity of the combustion model is investigated in the following using the simplified test case. The simplified test case is homogeneously initialised with $\phi = 2.0$, $T = 850$ K, $p = 20$ bar, and $\psi = 0.0$ and mixture transformation is initiated either by premixed flame propagation or auto-ignition. The different mesh sizes applied are listed in table A.1.

Name	Mesh Size
Fine Mesh	0.0005 m
Normal Mesh	0.0010 m
Coarse Mesh	0.0020 m

Table A.1: Mesh sizes applied for investigation of combustion model mesh size sensitivity

The resulting pressure curves and zeroth statistical soot moments are displayed in figure A.1.

For the G-equation model the pressure courses exhibit a mesh size dependency, whereat the combustion onset is shift towards earlier crank angle degrees with decreasing mesh size. The combustion onset is controlled by the flame kernel model, and combustion starts as soon as the actual kernel diameter exceeds a pre-defined critical value (see section 3.4). The accuracy of determining the actual kernel diameter and its ratio to the critical value increases with decreasing mesh size and combustion starts with a comparatively smaller (and more accurate) kernel diameter the smaller mesh size is. As a result, the total volume transformed by turbulent flame propagation increases with decreasing mesh size, which leads to an increase of the pressure gradient. The increased pressure gradient enhances the turbulent kinetic energy and thus scalar dissipation rate. Due to this, the maximum zeroth statistical soot moment reduces with decreasing mesh size.

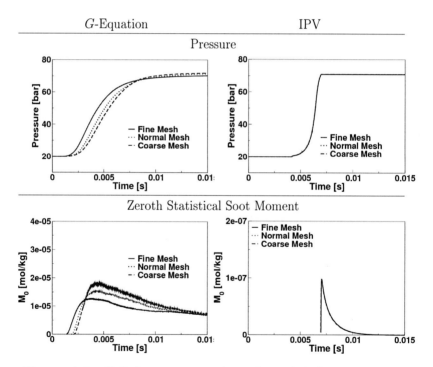

Figure A.1: Cylinder pressure and zeroth statistical soot moment calculated in simplified test case for mixture transformation initiated either by G-equation or IPV model at different mesh sizes and for $\phi = 2.0$, $T = 850$ K, $p = 20$ bar, and $\psi = 0.0$

In contrast to the G-equation model, the IPV model is not sensitive towards the mesh size.

A.1.2   Time Step Size Sensitivity

In the following, the time step size sensitivity of the combustion model is investigated using the simplified test case. The simplified test case is homogeneously initialised with $\phi = 2.0$, $T = 850$ K, $p = 20$ bar, and $\psi = 0.0$ and mixture transformation is initiated either by a spark plug or auto-ignition. The different time step sizes applied are listed in table A.2.

Name	Time Step Size
Half Time Step	0.0078125°CA
Normal Time Step	0.0156250°CA
Double Time Step	0.0312500°CA

Table A.2: Time step sizes applied for investigation of combustion model time step size sensitivity

The resulting pressure curves and zeroth statistical soot moments are displayed in figure A.2.

For the G-equation model the pressure courses show no sensitivity towards the time step size. However, the courses of the zeroth statistical soot moments exhibit an implausible solution for a large time step size as a result of stability constraints of explicate numerical procedures (for detailed description of stiff differential equations see [49]).

In contrary, the IPV model shows a time step size dependency, whereat the combustion onset is shift towards later crank angle degrees with decreasing time step size. The combustion onset is controlled by the development of the combustion progress variable, which is calculated at each time step as the sum of the progress variable at the old time step and the progress variable at the new time step. Latter one is defined as the product of the change of the progress variable at the new time step and the time step size (see section 4.4). Thus, a decrease of the time step size results in a decrease of the progress variable at the new time step and therefore in a later (and more accurate) combustion onset. The maximum values of the zeroth statistical soot moment are thereby only marginally affected.

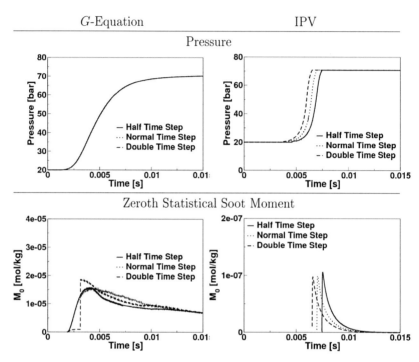

Figure A.2: Cylinder pressure and zeroth statistical soot moment calculated in simplified test case for mixture transformation initiated either by G-equation or IPV model at different time step sizes and for $\phi = 2.0$, $T = 850$ K, $p = 20$ bar, and $\psi = 0.0$

A.2   Highly Turbocharged SI Engine

A.2.1   Engine Specification

The highly turbocharged SI engine investigated bases on an 1.4l 4-cylinder engine. The engine has a supercharger and an exhaust gas turbocharger serially connected. Details about the charging concept can be found in [94, 95]. At low engine speeds the compressor can be used in addition to the exhaust gas turbocharger to enhance the boost pressure. Thus, high boost pressures and indicated mean effective pressures can be achieved. Additionally, the engine consists of a tumble flap located in the intake port, which can be activated to enhance the in-cylinder charge motion.

The specifications of the highly turbocharged SI engine are listed in table A.3.

Characteristic	Description
Type	4-cylinder in-line engine
Operating method	4-stroke SI engine
Swept volume	1390 cm^3
Bore/ Stroke	76.5 mm/ 75.6 mm
Compression ratio	9.5
Power rating	132 kW
Rated speed	6000 1/min
Maximal torque	250 Nm
Maximal mean effective pressure	23 bar

Table A.3: Specifications of the highly turbocharged SI engine

A.2.2 Operating Points

Investigations are carried out on two operating points, which are listed in table A.4.

Operating point	Engine speed [rpm]	BMEP [bar]	ϕ [-]	ψ [-]	Tumble flap
part load	2000	4.99	0.9/1.0/1.3	0.08	op.
full load	1500	20.17	1.0	0.02	op./cl.

Table A.4: Operating points and variations of the highly turbocharged SI engine

A.2.3 Chemiluminescence Measurements

The flame propagation characterisation is carried out using optical measurements. Herein, the electromagnetic radiations characterised by different intensities and electromagnetic spectra resulting from the combustion of hydrocarbon fuel are utilised [142]. The molecules emission and absorption levels differ and are dependent on the wavelength.

Electromagnetic radiation occurs when an electron changes its state from the state of lowest energy, the stationary state, to a state of higher energy, the excited state, by introducing the energy required [149]. When the excited electron changes its state back to the ground state, energy in form of light is emitted. In detail, the emitted light are photons of a specific wavelength. In case the excited state is a result of a chemical reaction, the electromagnetic radiation is called chemiluminescence [151].

Species possessing chemiluminescence during the combustion of a hydrocarbon fuel are CH, OH, H_2O, CH_2O and CO_2 [142]. Figure A.3 depicts the normalised mole fractions of the chemiluminescence species in-line with the temperature curve of a flame as a function of distance.

The flame temperature increases slightly until a distance of 1.2 mm. This increase is a result of pre-reactions associated with the so-called cool flame. Thereafter, a steep temperature gradient occurs which can be related to the main heat release.

The concentration of the species CH_2O peaks in the cool flame region. Since H_2O and CO_2 are combustion products, their concentrations increase as soon as the mixture transformation starts. Due to the pre-reactions occurring indicated by the enhanced CH_2O concentration, an increase of both species concentrations can be observed starting in the cool flame region. The radicals CH and OH peak in the range of the main heat release.

Utilising the chemiluminescence and keeping in mind that the species emission and absorption levels differ and are dependent on the wavelength, the flame can be classified in cool flame and main heat release using corresponding narrow and broad band filter.

In order to characterise the mean flame propagation, the OH radical chemiluminscence is detected. The radical has an electromagnetic spectrum in the range of $260.8 - 347.2$ nm, with a maximum value at 306.4 nm [142]. For this reason an optical band-pass filter in the range of $280.0 - 380.0$ nm is used. Chemiluminescence of other radicals occurs in the range of band-pass filter, too, i.e. CH with a maximum peak value in this range at 390 nm [42]. Additionally, a broad base emission line of CO, CO_2 and HCO can be observed [42, 122].

To enable optical in-cylinder measurements, an optical access in the engine cylinder head is required. Thereby, the specific measurement technique requirements need to be satisfied. Additionally, the optical access needs to be resistant to the high in-cylinder pressure and temperature values occurring during combustion. Both requirements can be fulfilled using the endoscope technique. The construction design of the endoscope access in the cylinder head is illustrated in figure A.4.

The chemiluminescence signals of the propagating flame front are recorded using a high speed camera allowing a resolution at 2000 rpm of every second crank angle degree. In order to increase the chemiluminescence signals a high speed image intensifier is used.

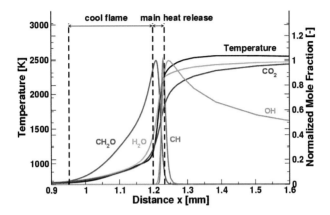

Figure A.3: Calculated concentrations of species possessing chemiluminescence during hydrocarbon combustion in a stoichiometric *iso*-Octane/air flame

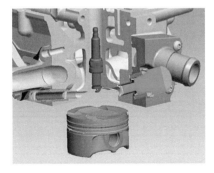

Figure A.4: Schematic illustration of endoscope access in cylinder head of the highly turbocharged SI engine

A.3 Small Mildly Turbocharged SI Engine

A.3.1 Engine Specification

The small mildly turbocharged SI engine investigated bases on an 1.4l 4-cylinder engine, too. The specifications of the engine are listed in table A.5.

Characteristic	Description
Type	4-cylinder in-line engine
Operating method	4-stroke SI engine
Swept volume	1390 cm^3
Bore/ Stroke	76.5 mm/ 75.6 mm
Compression ratio	10
Power rating	90 kW
Maximal torque	200 Nm

Table A.5: Specifications of the 1.4l mildly turbocharged SI engine [177]

A.3.2 Operating Point

Investigations are carried out at one operating point. Details about the operating point can be found in table A.6.

Engine speed [rpm]	BMEP [bar]	ϕ [-]	ψ [-]
1500	5.0	1.0	0.09

Table A.6: Operating point of the small mildly turbocharged SI engine

A.4 Big Mildly Turbocharged SI Engine

A.4.1 Engine Specification

The big mildly turbocharged SI engine investigated bases on an 1.8l 4-cylinder engine. The specifications of the engine are listed in table A.7.

Characteristic	Description
Type	4-cylinder in-line engine
Operating method	4-stroke SI engine
Swept volume	1798 cm^3
Bore/ Stroke	82.5 mm/ 84.2 mm
Compression ratio	9.6
Power rating	118 kW
Maximal torque	250 Nm

Table A.7: Specifications of the 1.8l mildly turbocharged SI engine [9]

A.4.2   Operating Points

Investigations are carried out on two full load operating points, which are listed in table A.8.

Operating point	Engine speed [rpm]	BMEP [bar]	ϕ [-]	ψ [-]
A	3000	17.47	1.0	0.02
B	5000	15.72	1.0	0.02

Table A.8: Operating points of the big mildly turbocharged SI engine

A.5   HCCI Engine

A.5.1   Engine Specification

The HCCI engine investigated is the Volkswagen Gasoline Compression Ignition (GCI) engine operating in the part load range on the bases of the homogeneous auto-ignition in combination with spark ignition. In order to initiate gasoline auto-ignition, a high temperature level is required in the end of the compression stroke. An effective way to achieve a temperature enhancement is to use the exhaust gases from the previous combustion cycle [187]. In general this can be done based on two different gas exchange strategies [26]: the exhaust gas retainment with early exhaust valves closing and late inlet valves opening (and intermediate compression) and the exhaust gas retraction from the outlet with a systematically repeated opening of the exhaust valves in the intake phase. Latter strategy is characterised by a decreased fuel consumption, increased running smoothness and extended characteristic engine map as well as robustness in terms of external impact parameters, and schematically illustrated in figure A.5.

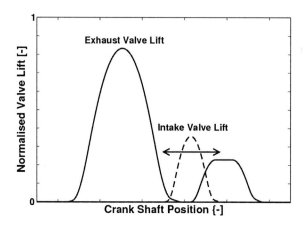

Figure A.5: Valve lifts applying exhaust gas retraction from the outlet as gas exchange strategy [187]

The GCI combustion process is applied to a 2.0l turbo engine which bases on the 2.0l-147kW-TFSI engine [187]. The specifications are listed in table A.9.

Characteristic	Description
Type	4-cylinder in-line engine
Operating method	4-stroke HCCI/SI engine
Swept volume	1984 cm^3
Bore/ Stroke	82.5 mm/ 92.8 mm
Compression ratio	10.0

Table A.9: Specifications of the HCCI engine

A.5.2   Operating Points

Based on the test bed measurements three steady-state ran operating points are considered in numerical investigations, which are listed in table A.10.

Operating point	Engine speed [rpm]	BMEP [bar]	ϕ [-]	ψ [-]	$\sigma_{p_{max}}$ [-]
A	1000	5.055	0.885	0.24	3.49
B	2000	4.031	0.826	0.38	2.78
C	2500	4.082	0.877	0.36	2.78

Table A.10: Investigated operating points of the HCCI engine

A.5.3   Cyclic Variations of Operating Point A

The 200 measured cycles of the HCCI engine at operating point A are displayed in figure A.6 in terms of 50 % transformation point $x_b = 0.5$ and combustion duration between $0.1 \leq x_b \leq 0.9$.

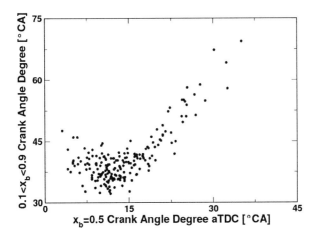

Figure A.6: 50 % transformation point and combustion duration of the 200 measured cycles of the HCCI engine at operating point A

A.6 Optical SI Engine

A.6.1 Engine Specification

The optical SI engine investigated consists of a single cylinder. For optical access, the liner is made of quartz glass. Optical measurements are carried out using high-speed-video equipment without additional lightening. Additionally, at the end of the exhaust gas system the engine is equipped with a particle counter from AVL company, type APC 489. The specifications of the optical engine are listed in table A.11.

Characteristic	Description
Type	1-cylinder engine
Operating method	4-stroke SI engine
Swept volume	349 cm^3
Bore/ Stroke	74.5 mm/ 80.0 mm
Compression ratio	8.02

Table A.11: Specifications of the optical SI engine

A.6.2 Operating Point

Investigations are carried out on a SOI variation at one operating point. Details about the operating point can be found in table A.12.

Engine speed [rpm]	Manifold Pressure [bar]	ϕ [-]	ψ [-]	SOI [°CA bTDC]
2200	0.73	1.0	0.21/0.19	248/328

Table A.12: Operating point and variation of the optical SI engine

A.6.3 Impact of Compression Ratio on Soot Formation

As outlined in section 5.9, the compression ratio of the optical SI engines is lower than the one of the complete engine due to omission of piston rings and thus increase of the compression volume. Typically, by modelling engines an increase of compression volume is taken into account by adjusting the piston position. The impact of piston position adjustment on the soot formation process is investigated in the following.

For this, calculations for a SOI at 330 °CA bTDC are carried out for both compression ratios, the one of the complete engine, 9.3, and the one of the optical engine, 8.02. Figure A.7 displays the liquid fuel injection and wall film formation for the two different compression ratios.

Compression ratio 9.3	Compression ratio 8.02
Liquid Fuel	

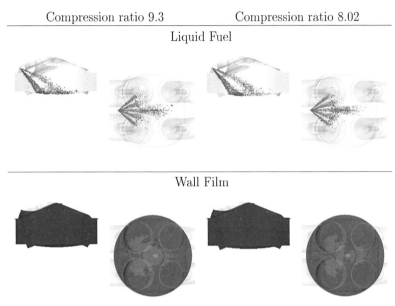

Wall Film

Figure A.7: Calculated liquid fuel injection (in side and top view at 315 °CA bTDC) and wall film formation (in side and top view at 305 °CA bTDC) for a SOI variation of the optical SI engine at part load operating point (2200 rpm, 0.73 bar) under varying compression ratios

For both compression ratios, the liquid fuel impinges directly on the piston resulting in wall film formation. However, for a compression ratio of 8.02, the mean free path of the liquid spray is longer than the one for a compression ratio of 9.3 due to piston position farther from TDC. As a result, with decreasing compression ratio the area covered with wall film is shift from the intake valve side/center in direction of the exhaust valve side.

In figure A.8 the resulting pressure curves of the two compression ratios are displayed.

The figure shows, that a decrease of compression ratio results in lower in-cylinder pressure values and a shift of maximal in-cylinder pressure towards later crank angle degree.

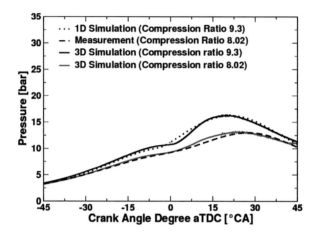

Figure A.8: Calculated and measured pressure curves for a SOI at 330 °CA bTDC of the optical SI engine at part load operating point (2200 rpm, 0.73 bar). Calculations are carried out for both, the compression ratio of the optical engine and the one of the complete engine.

Figure A.9 compares the evolution of wall film and soot during the combustion process for the two different compression ratios investigated.

For a compression ratio of 9.3, the wall film area decreases considerably in the period 10 °CA aTDC to 20 °CA aTDC. The largest area covered with soot occurs at 20 °CA aTDC. Whereat, the soot spreads from the center of the bowl into the center of the combustion chamber. A decrease of compression ratio to 8.02 retards the combustion process. Due to this, the wall film area decreases considerably at later crank angle degree, in the period 20 °CA aTDC to 30 °CA aTDC. The largest area covered with soot occurs at 40 °CA aTDC in the center of the bowl on the exhaust valve side[1].

---

[1]In contrast to the compression ratio of 9.3, the time period between wall film evaporation and soot formation is for the compression ratio of 8.02 increased. This is due to the decreased in-cylinder pressure value retarding the soot formation process.

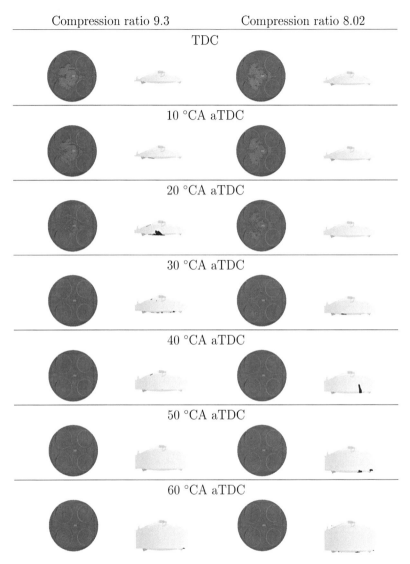

Figure A.9: Calculated in-cylinder distribution of wall film and soot for a SOI at 330 °CA bTDC of the optical SI engine at part load operating point (2200 rpm, 0.73 bar) for varying compression ratio. Soot iso-surface displayed corresponds to a soot mass of 1E-14 kg.

Compared with the optical measurements displayed in figure 5.26, the calculation carried out with the compression ratio of the optical engine predicts the correct soot formation time, but the wrong location. In contrast, the calculation based complete engine compression ratio predicts the correct soot formation location, but a too early soot formation onset.

Bibliography

[1] Abad Lozano, M. *Application of Genetic Algorithms for the Optimisation of Diesel Engine Combustion.* PhD Thesis, Karlsruhe University, 2006.

[2] Abraham, J., Williams, F.A., Bracco, F.V. *A Discussion of Turbulent Flame Structure in Premixed Charges.* SAE Technical Paper 850345, 1985.

[3] Aceves, S.M., Flowers, D.L., Martinez-Frias, J., Smith, J.R., Westbrook, C.K., Pitz, W.J., Dibble, R., Wright, J.F., Akinyemi, W.C., Hessel, R.P. *A Sequential Fluid-Mechanic Chemical-Kinetic Model of Propane HCCI Combustion.* SAE Technical Paper 2001-01-1027, 2001.

[4] Aceves, S.M., Flowers, D.L., Westbrook, C.K., Smith, J.R., Pitz, W., Dibble, R., Christensen, M., Johansson, B. *A Multi-Zone Model for Prediction of HCCI Combustion and Emissions.* SAE Technical Paper 2000-01-0327, 2000.

[5] Adomeit, P., Sehr, A., Weinowski, R., Stapf, K.G., Seebach, D., Pischinger,S., Hoffmann, K., Abel, D., Fricke, F., Kleeberg, H., Tomazic, D. *Operation Strategies for Controlled Auto Ignition Gasoline Engines.* SAE Technical Paper 2009-01-0300, 2009.

[6] Ahmed, S.S., Moréac, G., Zeuch, T., Mauß, F. *Reduced Mechanism for the Oxidation of the Mixtures of n-Heptane and iso-Octane.* Proceedings of the European Combustion Meeting "ECM 2005", Louvain-la-Neuve, Belgium, April 3-6, 2005.

[7] Ahmed, S.S. *A Detailed Modeling Study for Primary Reference Fuels and Fuel Mixtures and Their use in Engineering Applications.* PhD Thesis, Lund University, 2006.

[8] Aldredge, R.C., Vaezi, V., Ronney, P.D. *Premixed-flame propagation in turbulent Taylor-couette flow.* Combustion and Flame, Vol. 115, Issue 3, November 1998, pp. 395-405.

[9] Audi AG Service-Training *Self-Stufy Programme 384. Audi Chain-Driven 1.8 litre 4V TFSI engine*. A06.5S00.29.20, Technical Status August 2006.

[10] Bai, C.,Gosman, A.D. *Development of Methodology for Spray Impingement Simulation*. SAE Technical Paper 950283, 1995.

[11] Bai, C.,Gosman, A.D. *Mathematical modelling of wall films formed by impinging sprays*. SAE Technical Paper 960626, 1996.

[12] Barths, H., Antoni, C., Peters, N. *Three-Dimensional Simulation of Pollutant Formation in a DI Diesel Engine using Multiple Interactive Flamelets*. SAE Technical Paper 982459, 1998.

[13] Barths, H., Hasse, C., Peters, N. *Computational fluid dynamics modelling of non-premixed combustion in direct injection diesel engines*. International Journal of Engine Research, Vol. 1, No. 3, June 2000, pp. 249-267.

[14] Basshuysen, R.v., Schäfer, F. *Handbuch Verbrennungsmotor: Grundlagen, Komponenten, Systeme, Perspektiven*. Vieweg & Teubner, 6th edition, 2012.

[15] Beck, L.M. *Reaktionskinetische Analyse des Einflusses von chemisch modifizierter Abgasrückführung auf das Selbstzündungsverhalten von Ottokraftstoffen*. Diploma Thesis, Brandenburg University of Technology Cottbus, 2008.

[16] Beck, L.M., Mauß, F., Montefrancesco, E., Winkler, A. *Selbstzündung und Flammenausbreitung - Modellierungsansätze in 3D-CFD am Beispiel des Volkswagen GCI® Motors*. Virtual Powertrain Creation, 13th International MTZ conference, Unterschleißheim near Munich, December 8-9, 2011.

[17] Bédat, B., Cheng, R.K. *Experimental Study of Premixed Flames in Intense Isotropic Turbulence*. Combustion and Flame, Vol. 100, Issue 3, February 1995, pp. 485-494.

[18] Birkigt, A. *Measured Pre-Ignition Locations of a Tumble Flap Variation for a Highly Turbocharged SI Engine using an Optical 40-Channel Waveguide*. Interal Report, unpublished, 2010.

[19] Birkigt, A. *Analyse von Vorentflammungsphänomenen an hoch aufgeladenen Ottomotoren mit Direkteinspritzung*. PhD Thesis, Erlangen-Nürnberg University, 2011.

[20] Blunsdon, C.A., Dent, J.C. *The Simulation of Autoignition and Knock in a Spark Ignition Engine with Disk Geometry.* SAE Technical Paper 940524, 1994.

[21] Bo, T., Mauß, F., Beck, L.M. *Detailed Chemistry CFD Engine Combustion Solution with Ignition Progress Variable Library Approach.* SAE Technical Paper 2009-01-1898, 2009.

[22] Borman G.L., Johnson, J.H. *Unsteady Vaporization Histories and Trajectories of Fuel Drops Injected into Swirling Air.* SAE Technical Paper 620271, 1962.

[23] Boussinesq, J. *Théorie de l'éculement tourbillant.* Mémoires presents par Divers Savants Sciences Math. et Phys., Academie de Sciences, Paris, Vol. 23, 1877, pp. 46-50.

[24] Bradley, D., Merdjani, S., Sheppard, C.G.W., Yeo, J. *A Computational Model of Autoignition in Spark Ignition Engines.* Joint Meeting of the Portuguese, British, Spanish and Swedish Sections of the The Combustion Institute, Vol. 17, Issue 3, Funchal Madeira, April 1996, pp. 1-4.

[25] Bray, K.N.C., Libby, P.A. *Passage Times and Flamelet Crossing Frequencies in Premixed Turbulent Combustion.* Combustion Science and Technology, Vol. 47, Issue 5-6, 1986, pp. 253-274.

[26] Bücker, C., Stapf, G., Krebber-Hortmann, K., Pischinger, S., Mori, S. *Unterschiedliche Strategien für die Ventilsteuerung bei kontrollierter Selbstzündung CAI.* Conference on "Variable Ventilsteuerung", Essen, February 2007.

[27] Cant, R.S., Pope, S.B., Bray, K.N.C. *Modelling of Flamelet Surface-to-volume Ratio in Turbulent Premixed Combustion.* Symposium (International) on Combustion, Vol. 23, Issue 1, 1991, pp. 809-815.

[28] Cantrell, B.A., Ge, H.-W., Reitz, R.D., Rutland, C.J. *Validation of Advanced Combustion Models Applied to Two-Stage Combustion in a Heavy Duty Diesel Engine.* SAE Technical Paper 2009-01-0714, 2009.

[29] Chen, C. *Personal communication.* 2011.

[30] Cheng, R.K., Shepherd, I.G., Talbot, L. *Reaction Rates in Premixed Turbulent Flames and Their Relevance to the Turbulent Burning*

Speed. Symposium (International) on Combustion, Vol. 22, Issue 1, 1989, pp. 771-780.

[31] Cheng, R.K., Shepherd, I.G. *The Influence of Burner Geometry on Premixed Turbulent Flame Propagation.* Combustion and Flame, Vol. 85, Issue 1-2, May 1991, pp. 7-26.

[32] Cheng, W.K., Diringer, J.A. *Numerical Modelling of SI Engine Combustion with a Flame Sheet Model.* SAE Technical Paper 910268, 1991.

[33] Chin, J.S., Lefebvre, A.H. *The Role of the Heat-up Period in Fuel Drop Evaporation.* International Journal of Turbo and Jet Engines, Vol. 2, Issue 4, 1985, pp. 315-326.

[34] Choi, C.R., Huh, K.Y. *Development of a Coherent Flamelet Model for a Spark Ignited Turbulent Premixed Flame in a Closed Vessel.* Combustion and Flame, Vol. 114, Issue 3-4, 1998, pp. 336-348.

[35] Clavin, P., Williams, F.A. *Theory of Premixed-flame Propagation in Large Scale Turbulence.* Journal of Fluid Mechanics, Vol. 90, Issue 3, January 1979, pp. 589-604.

[36] Colin, O., Benkenida, A. *The 3-Zones Extended Coherent Flame Model (Ecfm3z) for Computing Premixed/Diffusion Combustion.* Oil and Gas Science and Technology, Vol. 59, Issue 6, 2004, pp. 593-609.

[37] Contino,F., Jeanmart,H. *Study of the HCCI Running Zone using Ethyl Acetate.* SAE Technical Paper 2009-01-0297, 2009.

[38] Crowe, C.T., Sharma, M.P., Stock, D.E. *The Particle-Source-in-Cell Method for Gas Droplet Flow.* ASME, Transactions, Series I - Journal of Fluids Engineering, Vol. 99, June 1977, pp. 325-332.

[39] Curran, H.J., Pitz, W.J., Westbrook, C.K., Callahan, C.V., Dryer, F.L. *Oxidation of Automotive Primary Reference Fuels at Elevated Pressures.* Symposium (International) on Combustion, Vol. 1, 1998, pp. 379-387.

[40] Damköhler, G. *Der Einfluß der Turbulenz auf die Flammengeschwindigkeit in Gasgemischen.* Zeitschrift für Elektrochemie und angewandte physikalische Chemie, Vol. 46, Issue 11, Nov. 1940, pp. 601-626.

[41] Davis, S.G., Law, C.K. *Laminar Flame Speeds and Oxidation Kinetics of iso-Octane-Air and n-Heptane-Air Flames.* Symposium (International) on Combustion, Vol. 1, 1998, pp. 521-527.

[42] Dec, J.E., Espey, C. *Chemiluminescence Imaging of Autoignition in a DI Diesel Engine.* SAE Technical Paper 982685, 1998.

[43] Dekena, M. *Numerische Simulation der turbulenten Flammenausbreitung in einem direkt einspritzenden Benzinmotor mit einem Flamelet-Modell.* PhD Thesis, Aachen University, 1998.

[44] Duclos, J.M., Veynante, D., Poinsot, T. *A Comparison of Flamelet Models for Premixed Turbulent Combustion.* Combustion and Flame, Vol. 95, Issue 1-2, October 1993, pp. 101-117.

[45] Duclos, J.M., Zolver, M. *3D Modeling of Intake, Injection and Combustion in DI-SI Engine under Homogeneous and Stratified Operating Conditions.* International Symposium on Diagnostics, Modelling of Combustion in Internal Combustion Engines (COMODIA), Vol. 4, 1998, pp. 335-340.

[46] Eckert, P., Kong, S.-C., Reitz, R.D. *Modeling Autoignition and Engine Knock Under Spark Ignition Conditions.* SAE Technical Paper 2003-01-0011, 2003.

[47] Egolfopoulos, F.N., Cho, P., Law, C.K. *Laminar Flame Speeds of Methane-air Mixtures under Reduced and Elevated Pressures.* Combustion and Flame, Vol. 76, Issue 3-4, June 1989, pp. 375-391.

[48] Embouazza, M., Haworth, D.C., Darabiha, N. *Implementation of Detailed Chemical Mechanisms into Multidimensional CFD using in situ Adaptive Tabulation: Application to HCCI Engines.* SAE Technical Paper 2002-01-2773, 2002.

[49] Engeln-Müllges, G., Niederdrenk, K., Wodicka, R. *Numerik-Algorithmen: Verfahren, Beispiele, Anwendungen.* 10th Edition, Springer, Berlin, 2011.

[50] Frenklach, M., Clary, D.W., Gardiner Jr., W.C., Stein, S.E. *Detailed Kinetic Modeling of Soot Formation in Shock-tube Pyrolysis of Acetylene.* Symposium (International) on Combustion, Vol. 20, Issue 1, 1985, pp. 887-901.

[51] Frenklach, M., Wang, H. *Detailed Modeling of Soot Particle Nu-cleation and Growth.* Symposium (International) on Combustion, Vol. 23, Issue 1, 1991, pp. 1559-1566.

[52] Frenklach, M. *Method of Moments with Interpolative Closure.* Chemical Engineering Science, Vol. 57, Issue 12, June 2002, pp. 2229-2239.

[53] Fröhlich, J. *Large Eddy Simulation turbulenter Strömungen.* B.G. Teubner, Wiesbaden, 2006.

[54] Gardiner Jr., W.C. *Gas-Phase Combustion Chemistry.* Springer New York, January 2000.

[55] Gelbard, F., Tambour, Y., Seinfeld, J.H. *Sectional Representations for Simulating Aerosol Dynamics.* Journal of Colloid and Interface Science, Vol. 76, Issue 2, 1980, pp. 541-556.

[56] Glassmann, I. *Combustion.* Academic Press Inc., 2nd Edition, 1987.

[57] Golloch, R. *Downsizing bei Ottomotoren: Ein wirkungsvolles Konzept zur Kraftstoffverbrauchssenkung.* Springer, Berlin, 2005.

[58] Golovitchev, V.I., Montorsi, L., Calik, A.T., Ergeneman, A.M. *Application of dynamic ϕ-T parametric maps to 3D detailed chemistry combustion analysis in diesel engines.* BEV Engine Combustion Processes, Current Problems and modern techniques, VIII congress, Munich, 2007.

[59] Golovitchev, V.I., Calik, A.T., Montorsi, L. *Analysis of Combustion Regimes in Compression Ignited Engines using Parametric ϕ-T Dynamic Maps.* SAE Technical Paper 2007-01-1838, 2007.

[60] Göttgens, J., Mauß, F., Peters, N. *Analytic Approximation of Burning Velocities and Flame Thicknesses of Lean Hydrogen, Methane, Ethylene, Ethane, and Propane Flames.* Symposium (International) on Combustion, Vol. 24, Issue 1, 1992, pp. 129-135.

[61] Gülder, Ö.L. *Correlations of Laminar Combustion Data for Alternative S.I. Engine Fuels.* SAE Technical Paper 841000, 1984.

[62] Gülder, Ö.L. *Turbulent Premixed Flame Propagation Models for Different Combustion Regimes.* Symposium (International) on Combustion, Vol. 23, Issue 1, 1991, pp. 743-750.

[63] Hadler, J. *Mobility in the conflict area of global energy chains.* 32nd International "Wiener Motoren Symposium", Vienna, 2011.

[64] Halstead, M.P., Kirsch, L.J., Quinn, C.P. *The Autoignition of Hydrocarbon Fuels at High Temperatures and Pressures - Fitting of a Mathematical Model.* Combustion and Flame, Vol. 30, Issue C, 1977, pp. 45-60.

[65] Hasse, C. *A Two-Dimensional Flamelet Model for Multiple Injections in Diesel Engines.* PhD Thesis, Aachen University, 2004.

[66] Heimel, S., Weast, R.C. *Effect of Initial Mixture Temperature on the Burning Velocity of Benzene-Air, n-Heptane-Air, and Iso-octane-Air Mixtures.* Symposium (International) on Combustion, Vol. 6, Issue 1, 1957, pp. 296-302.

[67] Hermann, A. *Modellbildung für die 3D-Simulation der Gemischbildung und Verbrennung in Ottomotoren mit Benzin-Direkteinspritzung.* PhD Thesis, Karlsruhe University, 2008.

[68] Herweg, R., Maly, R.R. *A Fundamental Model for Flame Kernel Formation in S. I. Engines.* SAE Technical Paper 922243, 1992.

[69] Herwig, H. *Strömungsmechanik - Einführung in die Physik von technischen Strömungen.* Vieweg and Teubner, Wiesbaden, 2008.

[70] Heywood, J.B. *Internal Combustion Engine Fundamentals.* McGraw-Hill Inc., 1988.

[71] Hiroyasu, H., Kadota, T. *Models for Combustion and Formation of Nitric Oxide and Soot in Direct Injection Diesel Engines.* SAE Technical Paper 760129, 1976.

[72] Hong, S., Wooldrige, M.S., Im, H.G., Assanis, D.N., Pitsch, H. *Development and Application of a Comprehensive Soot Model for 3D CFD Reacting Flow Studies in a Diesel Engine.* Combustion and Flame, Vol. 143, Issue 1-2, October 2005, pp. 11-26.

[73] Hu, D. *Modellierung und Modellentwicklung der Rußbildung bei hohem Druck in vorgemischten Verbrennungssystemen.* PhD Theses, Stuttgart University, 2001.

[74] James, E.H. *Laminar burning velocities of iso-octane-air mixtures - A literature review.* SAE Technical Paper 870170, 1987.

[75] Jerzembeck, S., Peters, N., Pepiot-Desjardins, P., Pitsch, H. *Laminar Burning Velocities at High Pressure for Primary Reference Fuels and Gasoline: Experimental and Numerical Investigation.* Combustion and Flame, Vol. 156, Issue 2, February 2009, pp. 292-301.

[76] Jia, M., Xie, M., Peng, Z. *Prediction of the Operating Range for a HCCI Engine Based on a Multi-zone Model.* SAE Technial Paper 2008-01-1663, 2008.

[77] Joos, F. *Technische Verbrennung. Verbrennungstechnik, Verbrennungsmodellierung, Emissionen.* Springer-Verlag, Berlin, Heidelberg, 2006.

[78] Juneja, H., Sczomak, D.P., Ge, H.-W., Yang, S., Reitz, R.D. *Application of a G-Equation Based Combustion Model and Detailed Chemistry to Prediction of Autoignition in a Gasoline Direct Injection Engine.* Proceedings of the 8th Congress on Gasoline Direct Injection Engines, Augsburg, September 2009.

[79] Kamimoto, T., Bae, M. *High Combustion Temperature for the Reduction of Particulate in Diesel Engines.* SAE Technical Paper 880423, 1988.

[80] Karlsson, A., Magnusson, I., Balthasar, M., Mauß, F. *Simulation of Soot Formation under Diesel Engine Conditions using a Detailed Kinetic Soot Model.* SAE Technical Paper 981022, 1998.

[81] Kawanabe, H., Shioji, M., Tsunooka, T., Ali, Y. *CFD Simulation for Predicting Combustion and Pollutant Formation in a Homogeneous-Charge Spark Ignition Engine.* The Fourth International Symposium COMODIA 98, 1998, pp. 233-238.

[82] Kazakov, A., Foster, D.E. *Modeling of Soot Formation During DI Diesel Combustion using a Multi-Step Phenomenological Model.* SAE Technical Paper 982463, 1998.

[83] Kennedy, I.M. *Models of Soot Formation and Oxidation.* Progress in Energy and Combustion Science, Vol. 23, Issue 2, 1997, pp. 95-132.

[84] Kieberger, M., Hofmann, P., Geringer, B., Thiele, F., Winkler, A., Frambourg, M. *Effects of Fuel Parameters in the Preignition Tendency of Highly Charged SI-Engines with Direct Injection.* 3rd conference "Ottomotorisches Klopfen - irreguläre Verbrennung", Berlin, 2010.

[85] Klimov, A.M. *Premixed Turbulent Flames - Interplay of Hydrodynamic and Chemical Phenomena.* Flames, Lasers and Reactive Systems, New York, American Institute of Aeronautics and Astronautics, Inc., 1983, p. 133-146.

[86] Knop, V., Jay, S. *Latest Developments in Gasoline Auto-Ignition Modelling applied to an optical CAI^{TM} Engine.* Oil and Gas Science and Technology, Vol. 61, 2006, pp. 121-138.

[87] Knop, V., Thirouard, B., Chérel, J. *Influence of the Local Mixture Characteristics on the Combustion Process in a CAIŹ Engine.* SAE Technical Paper 2008-01-1671, 2008.

[88] Kolla, H., Rogerson, J.W., Chakraborty, N., Swaminathan, N. *Scalar Dissipation Rate Modeling and its Validation.* Combustion Science and Technology, Vol. 181, 2009, pp. 518-535.

[89] Kolla,H., Rogerson, J.W., Swaminathan, N. *Validation of a Turbulent Flame Speed Model across Combustion Regimes.* Combustion Science and Technology, Vol. 182, 2012, pp. 284-308.

[90] Kolmogorov, A.N. *Integratable form of droplet drag coefficient.* Doklady Akademii Nauk SSSR, Vol. 30, 1941, pp. 299-303.

[91] Kong, S.-C., Han, Z., Reitz, R.D. *The Development and Application of a Diesel Ignition and Combustion Model for Multidimensional Engine Simulation.* SAE Technical Paper 950278, 1995.

[92] Kong, S.-C., Marriott, C.D., Reitz, R.D., Christensen, M. *Modeling and Experiments of HCCI Engine Combustion using Detailed Chemical Kinetics with Multidimensional CFD.* SAE Technical Paper 2001-01-1026, 2001.

[93] Kraus, E. *Simulation der vorgemischten Verbrennung in einem realen Motor mit dem Level-Set Ansatz.* PhD Thesis, Tübingen University, 2007.

[94] Krebs, R., Szengel, R., Middendorf, H., Fleiß, M., Laumann, A., Voeltz, S. *Neuer Ottomotor mit Direkteinspritzung und Doppelaufladung von Volkswagen - Teil 1: Konstruktive Gestaltung.* "Motortechnische Zeitschrift", Vol. 66, Issue 12, 2005.

[95] Krebs, R., Szengel, R., Middendorf, H., Sperling, H., Siebert, W., Theobald, J., Michels, K. *Neuer Ottomotor mit Direkteinspritzung*

und Doppelaufladung von Volkswagen - Teil 2: Thermodynamik. "Motortechnische Zeitschrift", Vol. 66, Issue 12, 2005.

[96] Kuo, K.K. *Principles of Combustion.* John Wiley & Sons, Inc., Hoboken, New Jersey, 2005.

[97] Lee, D., Han, I., Huh, K.Y., Lee, J.-H., Kim, S.-J., Kang, W., Kim, Y. *A New Combustion Model Based on Transport of Mean Reaction Progress Variable in a Spark Ignition Engine.* SAE Technial Paper 2008-01-0964, 2008.

[98] Lee, K., Min, K. *Study of a Stratification Effect on Engine Performance in Gasoline HCCI Combustion by using the Multi-zone Method and Reduced Kinetic Mechanism.* SAE Technical Paper 2009-01-1784, 2009.

[99] Lehtiniemi, H., Mauß, F., Balthasar, M., Magnusson, I. *Modeling Diesel Engine Combustion with Detailed Chemistry using a Progress Variable Approach.* SAE Technical Paper 2005-01-3855, 2005.

[100] Lehtiniemi, H., Mauß, F., Balthasar, M., Magnusson, I. *Modeling Diesel Spray Ignition using Detailed Chemistry with a Progress Variable Approach.* Combustion Science and Technology, Vol. 178, Issue 10-11, 2006, pp. 1977-1997.

[101] Li, G., Bo, T., Chen, C., Johns, R.J.R. *CFD Simulation of HCCI Combustion in a 2-Stroke DI Gasoline Engine.* SAE Technical Paper 2003-01-1855, 2003.

[102] Liang, L., Reitz, R.D. *Spark Ignition Engine Combustion Modeling using a Level Set Method with Detailed Chemistry.* SAE Technical Paper 2006-01-0243, 2006.

[103] Liang, L., Reitz, R.D., Iyer, C.O., Yi, J. *Modeling Knock in Spark-Ignition Engines using a G-equation Combustion Model Incorporating Detailed Chemical Kinetics.* SAE Technical Paper 2007-01-0165, 2007.

[104] Lipatnikov, A.N., Chomiak, J. *Turbulent Flame Speed and Thickness: Phenomenology, Evaluation, and Application in Multi-Dimensional Simulations.* Progress in Energy and Combustion Science, Vol. 28, Issue 1, 2002, pp. 1-74.

[105] Livengood, J.C., Wu, P.C. *Correlation of Autoignition Phenomena in Internal Combustion Engines and Rapid Compression Machines.* Symposium (International) on Combustion, Vol. 5, Issue 1, 1955, pp. 347-356.

[106] Lucas, G. *Influence of Geometry Modifications of a Combustion Chamber on Knock.* Master Thesis, University of Applied Sciences Esslingen, 2010.

[107] Magnussen, B.F., Hjertager, B.H. *On Mathematical Modeling of Turbulent Combustion with Special Emphasis on Soot Formation and Combustion.* Symposium (International) on Combustion, Vol. 16, Issue 1, 1977, pp. 719-729.

[108] Mantel, T., Borghi, R. *A New Model of Premixed Wrinkled Flame Propagation based on a Scalar Dissipation Equation.* Combustion and Flame, Vol. 96, Issue 4, 1994, p. 443.

[109] Marble, F.E., Broadwell, J.E. *The Coherent Flame Model for Turbulent Chemical Reactions.* Project Squid, Technical Report TRW-9-PU, Purdue University, 1977.

[110] Marquetand, J.O. *Simulation der Rußbildung in Flammen und Stoßrohren mit einem detaillierten und einem semi-empirischen Modell.* PhD Thesis, Heidelberg University, 2010.

[111] Matsumoto, S., Saito, S. *Monte Carlo Simulation of Horizontal Pneumatic Conveying Based on the Rough Wall Model.* Journal of Chemical Engineering of Japan, Vol. 3, 1970, pp. 223-230.

[112] Mattavi, J.N., Groff, E.G., Matekunas, F.V. *Turbulence, Flame Motion and Combustion Chamber Geometry - their Interactions in a Lean-Combustion Engine.* Proceedings of the Institute of Mechanical Engineers Conference on Fuel Economy and Emissions of Lean Burn Engine, Paper C100, 1979.

[113] Mauß, F., Schäfer, T., Bockhorn, H. *Inception and Growth of Soot Particles in Dependence on the Surrounding Gas Phase.* Combustion and Flame, Vol. 99, Issues 3-4, December 1994, pp. 697-705.

[114] Mauß, F. *Entwicklung eines kinetischen Modells der Rußbildung mit schneller Polymerisation.* PhD Thesis, Aachen University, 1998.

[115] Mauß, F., Keller, D., Peters, N. *A Lagrangian Simulation of Flamelet Extinction and Re-Ignition in Turbulent Jet Diffusion Flames.* Symposium (International) on Combustion, Vol. 23, Issue 1, 1991, pp. 693-698.

[116] Mauß, F., Netzell, K., Marchal, C., Moréac, G. *Modelling the Soot Particle Size Distribution Functions using a Detailed Kinetic Soot Model and a Sectional Method.* Combustion generated fine carbonaceous particles: Proceedings of an International Workshop held in Villa Orlandi, Anacapri, May 2007, pp. 465-482.

[117] Mauß, F., Ebenezer, N., Lehtiniemi, H. *Adaptive Polynomial Tabulation (APT): A computationally economical strategy for the HCCI engine simulation of complex fuels.* SAE Technical Paper 2010-01-1085, 2010.

[118] Merker, G., Schwarz, C., Stiesch, G., Otto, F. *Verbrennungsmotoren. Simulation der Verbrennung und Schadstoffbildung.* Vieweg and Teubner, 2nd Edition, 2007.

[119] Metghalchi, M., Keck, J.C. *Burning Velocities of Mixtures of Air with Methanol, Isooctane, and Indolene at High Pressure and Temperature.* Combustion and Flame, Vol. 48, 1982, pp. 191-210.

[120] Miller, J., Bowman, C.T. *Mechanism and Modeling of Nitrogen Chemistry in Combustion.* Progress in Energy and Combustion Science, Vol. 28, Issue 4, 1989, pp. 287-338.

[121] Montefrancesco, E. *Prediction of Premixed/Non-Premixed Combustion and Soot Formation using an Interactive Flamelet Approach.* PhD Thesis, University of Salento, 2007.

[122] Möser, P. *Zeitlich hochaufgelöste emissionsspektroskopische Untersuchung des Verbrennungsvorgangs im Otto-Motor.* PhD Thesis, Aachen University, 1994.

[123] Nagle, J., Strickland-Constable, R.F. Proceedings of the Fifth Conference on Carbon, Vol. 1, 1962, p. 154.

[124] Nakov, G., Mauß, F., Wenzel, P., Steiner, R., Krüger, C., Zhang, Y., Rawat, R., Borg, A., Perlman, C., Fröjd, K., Lehtiniemi, H. *Soot Simulation under Diesel Engine Conditions using a Flamelet Approach.* SAE Technical Paper 2009-01-2679, 2006.

[125] Netzell, K. *Development and Application of Detailed Kinetic Models for the Soot Particle Size Distribution Function.* PhD Thesis, Lund University, 2006.

[126] Netzell, K., Lehtiniemi, H., Mauß, F. *Calculating the Soot Particle Size Distribution Function in Turbulent Diffusion Flames using a Sectional Method.* Proceedings of the Combustion Institute, Vol. 31, Issue 1, January 2007, pp. 667-674.

[127] Nicholls, J.A. *Stream and Droplet Breakup by Shock Waves.* NASA SP-194, 1972.

[128] Nishida, K., Hiroyasu, H. *Simplified Three-Dimensional Modeling of Mixture Formation and Combustion in a D.I. Diesel Engine.* SAE Technical Paper 890269, 1989.

[129] O'Rourke, P.J., Amsden, A.A. *The TAB Method for Numerical Calculation of Spray Droplet Breakup.* SAE Technical Paper 872089, 1987.

[130] Patterson, M.A., Reitz, R.D. *Modeling the effect of fuel spray characteristics on diesel engine combustion and emission.* SAE Technical Paper 980131, 1998.

[131] Perlman, C. *Correlation function for laminar flame speed (iso-Octane)* LOGE internal report, 2010.

[132] Peters, N. *Turbulent Combustion.* Cambridge University Press, August 2000.

[133] Peters, N. *Laminar Flamelet Concepts in Turbulent Combustion.* Symposium (International) on Combustion, Vol. 21, Issue 1, 1988, pp. 1231-1250.

[134] Peters, N. *Fifteen Lectures on Laminar and Turbulent Combustion.* Ercoftac Summer School, Aachen, 1992.

[135] Peters, N. *Four Lectures on Turbulent Combustion.* Ercoftac Summer School, Aachen, 1997.

[136] Peters, N. *Technische Verbrennung.* Lecture notes, 2006.

[137] Pischinger, F., Spicher, U. *Spatial Flame Propagation and Flame Quenching During Combustion in Internal Combustion Engines.* SAE Technical Paper 845000, 1984.

[138] Pitsch, H., Wan, Y.P., Peters, N. *Numerical Investigation of Soot Formation and Oxidation Under Diesel Engine Conditions.* SAE Technical Paper 952357, 1995.

[139] Pocheau, A. *Scale Invariance in Turbulent Front Propagation.* Physical Review E, Vol. 49, Article 1109, 1994.

[140] Pope, S.B., Anand, M.S. *Flamelet and Distributed Combustion in Premixed Turbulent Flames.* Symposium (International) on Combustion, Vol. 20, Issue 1, 1985, pp. 403-410.

[141] Pope, S.B. *Computationally Efficient Implementation of Combustion Chemistry using In Situ Adaptive Tabulation.* Combustion Theory and Modelling, Vol. 1, 1997, pp. 44-63.

[142] Pöschl, M. *Einfluss von Temperaturinhomogenitäten auf den Reaktionsablauf bei der klopfenden Verbrennung.* PhD Thesis, Munich University, 2006.

[143] Putnam, A. *Integratable Form of Droplet Drag Coefficient.* American Rocket Society Journal, Vol. 31, 1961, pp. 1467-1468.

[144] Ranz, W.E., Marshall, W.R. *Evaporation from Drops, Part I.* Chemical Engineering Progress, Vol. 48, Issue 4, 1952, pp. 141-146.

[145] Reitz, R.D., Diwakar, R. *The Effect of Droplet Breakup on Fuel Sprays.* SAE Technical Paper 860469, 1986.

[146] Reitz, R.D. *Modeling Atomization Processes in High-Pressure Vaporizing Sprays.* Atomisation and Spray Technology, Vol. 3, Issue 4, 1987, pp. 309-337.

[147] Reitz, R.D., Diwakar, R. *Structure of High-Pressure Fuel Sprays.* SAE Technical Paper 870598, 1987.

[148] Ricart, L.M., Xin, J., Bower, G.R., Reitz, R.D. *In-cylinder Measurement and Modeling of Liquid Fuel Spray Penetration in a Heavy-Duty Diesel Engine.* SAE Technical Paper 971591, 1997.

[149] Riedel, E. *Allgemeine und anorganische Chemie.* deGruyter textbook, 7th edition, Berlin, 1999.

[150] Ricardo Software *VECTIS Manual.* 2008.

[151] Samaniego, J.-M., Egolfopoulos, F.N., Bowman, C.T. *CO_2^* Chemiluminescence in Premixed Flames.* Combustion Science and Technology, Vol. 109, Issue 1-6, 1995, pp. 183-203.

[152] Sauter, W., Hensel, S., Spicher, U., Schubert, A., Schießl, R., Maas, U. *Experimental and Numerical Investigation of Auto-Ignition Mechanisms for a Gasoline HCCI Mode.* 16th Aachener colloqium, 2007.

[153] Schießl, R., Maiwald, O., König, K., Maas, U. *Laserdiagnostische Untersuchung und detaillierte numerische Modellierung der Zündung in einem HCCI Motor.* 6th (International) Symposium on combustion diagnostics, Baden-Baden, 2004.

[154] Schlichting, H., Gersten, K. *Grenzschicht-Theorie.* Springer, 9th edition, 1997.

[155] Schmidt, H.-P., Habisreuther, P., Leuckel, W. *A Model for Calculating Heat Release in Premixed Turbulent Flames.* Combustion and Flame, Vol. 113, Issues 1-2, April 1998, pp. 79-91.

[156] Schreiber, M., Sadat Sakak, A., Poppe, C., Griffith, J.F., Halford-Maw, P., Rose, D.J. *Spatial Structure in End-Gas Autoignition.* SAE Technical Paper 932758, 1993.

[157] Seebach, D., Stapf, K.G., Pischinger, S., Hoffmann, K., Abel, D. *Modelling and Actuation of the Controlled Auto Ignition.* "Motorische Verbrennung: aktuelle Probleme und moderne Lösungsansätze", Energie- und Systemtechnik, 2009, pp. 395-406.

[158] Seidel, L. *Analyse des Einflusses der AGR-Rate und Zusammensetzung auf die Verbrennung in einem HCCI Motor.* Seminar Paper, Brandenburg University of Technology Cottbus, 2009.

[159] Senda, J., Kanda, T. *Modeling Spray Impingement Considering Fuel Film Formation on the Wall.* SAE Technical Paper 970047, 1997.

[160] Shepherd, I.G., Cheng, R.K. *The Burning Rate of Premixed Flames in Moderate and Intense Turbulence.* Combustion and Flame, Vol. 127, Issue 3, November 2001, pp. 2066-2075.

[161] Shuen, J.S., Chen, L.D., Faeth, G.M. *Evaluation of a Stochastic Model of Particle Dispersion in a Turbulent Round Jet.* AIChE Journal, Vol. 29, Issue 1, January 1983, pp. 167-170.

[162] Singh, S., Reitz, R.D., Wickman, D., Stanton, D., Tan, Z. *Development of a Hybrid, Auto-Ignition/Flame-Propagation Model and*

Validation Against Engine Experiments and Flame Liftoff. SAE Technical Paper 2007-01-0171, 2007.

[163] Spalding, D.B. *The Combustion of Liquid Fuels.* Fourth Symposium (International) on Combustion, Williams & Wilkins, Baltimore, 1953, pp. 847-864.

[164] Spalding, D.B. *Mixing and Chemical Reaction in Steady Confined Turbulent Flames.* Symposium (International) on Combustion, Vol. 13, Issue 1, 1971, pp. 649-657.

[165] Stanton, D.W., Rutland, C.J. *Modeling Fuel Film Formation and Wall Interaction in Diesel Engines.* SAE Technical Paper 960628, 1996.

[166] Stein, S., Budde, M., Krause, S., Brandt, S., Schlerege, F. *Beeinflussung der Schmierölemissionen durch die Gemischbildung im Brennraum von Verbrennungsmotoren.* FVV Project 933, Final Report, 2009.

[167] Steiner, R., Bauer, C., Krüger, C., Otto, F., Maas, U. *3D-Simulation of DI-Diesel Combustion Applying a Progress Variable Approach Accounting for Complex Chemistry.* SAE Technical Paper 2004-01-0106, 2004.

[168] Stow, C.D., Hadfield, M.G. *An Experimental Investigation of Fluid Flow Resulting from the Impact of a Water Drop with an Unyielding Dry Surface.* Proceedings of the Royal Society London, Vol. A 373, 1981, pp. 419-441.

[169] Strauss, T.S. *Simulation der Verbrennung und NO_x-Bildung in einem direktteinspritzenden Dieselmotor mit externer Abgasrückführung.* PhD Thesis, RWTH Aachen, 1998.

[170] Tabaczynski, R.J., Ferguson, C.R., Radhakrishnan, K. *A Turbulent Entrainment Model for Spark-Ignition Engine Combustion.* SAE Technical Paper 770647, 1977.

[171] Tan, Z., Kong, S.-C., Reitz, R.D. *Modeling Premixed and Direct Injection SI Engine Combustion using the G-Equation Model.* SAE Technical Paper 2003-01-1843, 2003.

[172] Tanner, F. *Liquid Jet Atomization and Droplet Breakup Modeling of Non-Evaporating Diesel Fuel Sprays.* SAE Technical Paper 970050, 1997.

[173] The European Parlament and the council of the european union *Directive 2009/28/EC of the European parlament and of the council of 23 April 2009 on the promotion of the use of energy from renewable sources and amending and subsequently repealing Directives 2001/77/EC and 2003/30/EC.* Official Journal of the European Union, L 140, 2009, pp. 16-62.

[174] Tominaga, R., Morimoto, S., Kawabata, Y., Matsuo, S., Amano, T. *Effects of Heterogeneous EGR on the Natural Gas Fueled HCCI Engine using Experiments, CFD and Detailed Kinetics.* SAE Technical Paper 2004-01-0945, 2004.

[175] Tzanetakis, T., Singh, P., Chen, J.-T., Thomson, M.J., Koch, C.R. *Knock Limit Prediction via Multi-Zone Modelling of a Primary Reference Fuel HCCI Engine.* International Journal of Vehicle Design, Vol. 54, Issue 1, 2010, pp. 47-72.

[176] Veynante, D., Vervisch, L. *Turbulent Combustion Modeling.* Progress in Energy and Combustion Science, Vol. 28, Issue 3, 2002, pp. 193-266.

[177] Volkswagen AG Service-Training *Self-Stufy Programme 405. 1.4l 90kW TSI Engine with Turbocharger.* 000.2812.05.20, Technical Status September 2007.

[178] Warnatz, J., Maas, U., Dibble, R.W. *Combustion. Physical and Chemical Fundamentals, Modeling and Simulation, Experiments, Pollutant Formation.* Springer-Verlag, 4th Ed., Berlin, Heidelberg, New York, 2006.

[179] Weber, J., Peters, N., Bockhorn, H., Pittermann, R. *Numerical Simulation of the Evolution of the Soot Particle Size Distribution in a DI Diesel Engine using an Emulsified Fuel of Diesel-Water.* SAE Technical Paper 2004-01-1840, 2004.

[180] Weber, J., Peters, N., Diwakar, R., Siewert, R.M., Lippert, A. *Simulation of the Low-Temperature Combustion in a Heavy Duty Diesel Engine.* SAE Technical Paper 2007-01-0904, 2007.

[181] Weller, H.G. *The Development of a new Flame Area Combustion Model using Conditional Averaging.* Thermo-Fluids Section Report TF/9307, Imperial College of Science Technology and Medicine, 1993.

[182] Weller, H.G., Uslu, S., Gosman, A.D., Maly, R.R., Herweg, R., Heel, B. *Prediction of Combustion in Homogeneous-Charge Spark-Ignition Engines.* International Symposium COMODIA 94, 1994.

[183] Wenzel, P., Gezgin, A., Steiner, R., Krüger, C., Netzell, K., Lehtiniemi, H., Mauß, F. *Modeling of the Soot Particle Size Distribution in Diesel Engines.* THIESEL, 2004.

[184] Westbrook, C.K., Pitz, W. *Detailed Kinetic Modeling of Autoignition Chemistry.* SAE Technical Paper 872107, 1987.

[185] Westbrook, C.K., Warnatz, J., Pitz, W.J. *A Detailed Chemical Kinetic Reaction Mechanism for the Oxidation of iso-Octane and n-Heptane over an Extended Temperature Range and its Application to Analysis of Engine Knock.* Proceedings of the Combustion Institute, Vol. 22, Issue 1, 1989, pp. 893-901.

[186] Willand, J. *Motortrends zur CO_2-Reduzierung: Downsizing oder Teillastbrennverfahren?* MTZ conference "Motor", Munich, June 03-04, 2008.

[187] Willand, J., Jakobs, J., Montefrancesco, E., Daniel, M., Vortkamp, V., Läer, B. *The Volkswagen GCI Combustion System for Gasoline Engines - Potentials and Limits in CO2 Emissions.* 30th International "Motorensymposium", Vienna, 2009.

[188] Willand, J., Daniel, M., Montefrancesco, E., Geringer, B., Hofman, P., Kieberger, M. *Grenzen des Downsizing bei Ottomotoren durch Vorentflammungen.* MTZ "Motortechnische Zeitschrift", Vol. 05, 2009.

[189] Williams, F.A. *Spray Combustion and Atomization.* Physics of Fluids, Vol. 1, Issue 6, 1958, pp. 541-545.

[190] Williams, F.A. *Turbulent combustion.* Frontiers in applied mathematics, The mathematics of combustion, Society of industrial and applied mathematics, Philadelphia, Pennsylvania, 1985.

[191] Wirth, M., Keller, P., Peters, N. *A Flamelet Model for Premixed Turbulent Combustion in SI-Engines.* SAE Technical Paper 932646, 1993.

[192] Yang, S., Reitz, R.D. *Improved combustion submodels for modeling gasoline engines with the level set G equation and detailed chemical kinetics.* Proceedings of the Institution of Mechanical Engineers,

Part D: Journal of Automobile Engineering, Vol. 223, Issue 5, May 2009, pp. 703-726.

[193] Yao, S.C., Cai, K.Y. *Dynamics and Heat Transfer of Drop Impacting on a Hot Surface at Small Angles.* ICLASS, 1985.

[194] Zeldovich, Y.B. *Acta Physicochimica.* USSR 21, 557, 1946.